The Conscious Universe

by

Tom Crawford

Bloomington, IN Milton Keynes, UK

AuthorHouse™
1663 Liberty Drive, Suite 200
Bloomington, IN 47403
www.authorhouse.com
Phone: 1-800-839-8640

AuthorHouse™ UK Ltd.
500 Avebury Boulevard
Central Milton Keynes, MK9 2BE
www.authorhouse.co.uk
Phone: 08001974150

© 2007 Tom Crawford. All rights reserved.

No part of this book may be reproduced, stored in a retrieval system, or transmitted by any means without the written permission of the author.

First published by AuthorHouse 11/7/2007

ISBN: 978-1-4343-0090-4 (sc)

Printed in the United States of America
Bloomington, Indiana

This book is printed on acid-free paper.

ABOUT THE AUTHOR

Tom Crawford PhD FRSE has held a lifelong interest in the link between philosophy and science. His research career began developing scientific measuring instruments and mathematical models. This is his first attempt at writing to his wider passion. His other enthusiasms include hill walking and flying light aircraft.

To Friends

Contents

about the author
preface
chapter 1
questions from everyday 1

chapter 2
the potter's kiln 13

chapter 3
down to earth 43

chapter 4
climbing up complexity 69

chapter 5
summit brain 112

chapter 6
the hidden outer world 138

chapter 7
limits to intuition 153

chapter 8
the unaccounted inner world 169

chapter 9
conscious machines 187

chapter 10
taking stock 197

post script

further reading list

PREFACE

My interest in the questions discussed in this book first developed in childhood. Like many others of my generation in Scotland, I played with that best of all scientist's or engineer's starter kits, the *'Meccano Set'*. This was a self-build toy of metal strips, wheels and imagination. From visually obvious, tactile, mechanical construction my schoolboy interests shifted to connecting up simple electrical circuits. This developed into building complete radio transmitters and receivers: a progression from objects which touched each other to objects which interacted at a distance. Copper wire wound on a cardboard tube: a few other components, resistors, capacitors, batteries, a single tube or transistor and a pair of headphones. I have never forgotten the wonder that surrounded such a simple apparatus. A piece of metal in my room, made to move by someone far off in another country. The question 'how does it do this?' led from construction to physics. And the question 'how do we know this?' led from physics to philosophy and later to the question 'why are things like this?'

Now I have tried to set out a more complete story of what emerges if we just keep asking what lies behind the most everyday experiences. This book contains reviews of specialist knowledge gained by countless people. Its contribution, I hope, is to link these areas of specialist knowledge within the framework of a single fundamental question.

We are conscious of seeing, hearing, smelling, touching, thinking and of pleasure and pain. Where does our conscious experience come from? How can we reconcile the scientific idea of us as being made up only of material structures, with our most central personal experience of a life full of sensations, thoughts and emotions? What happened in the universe to move it on, from objects with the modest complexity of a star, to objects with the immense complexity of a human brain?

CHAPTER I

questions from everyday

We live a miracle everyday, a world of coloured reflections in a river, scents of flowers, birdsong, a warm breeze. Buried below our living experience, science has discovered a world of atoms and vibrations, a world of blind elementary forces. These forces operate everywhere in the known universe, including the chemistry of our bodies. Most poignantly they operate in our brains. How does the miracle of our conscious life emerge from this dry world of atomic forces? Should we expect science to give a satisfactory answer to this question?

Science has done an amazing job in expanding our knowledge of the universe. It reveals new facts, features and connections, at an ever-accelerating rate. Building this body of knowledge constitutes one of the greatest triumphs of human intellect. However these achievements can blind us to how little we actually know about the world. Standing back, we can see that everything we observe has been filtered by our senses. There is a filter between what we perceive and what is going on in the external world which causes our perceptions. We try to understand our world in terms of what we observe and how we come to expect it to behave. Our conscious experience leads to ideas. Then ideas lead to explanations. With these explanations we now try to explain consciousness itself. Mind explains mind. Is this an exercise which is feasible? Is mind not the endpoint of the chain of all human explanation? Mind does the explaining!

Everything we know from science ties consciousness to elaborate nervous systems. We conclude that before life developed on Earth, consciousness did not exist on the planet.

So a path of rising complexity in the arrangement of material seems to precede the arrival of feeling and consciousness, from elementary atoms and molecules to simple organisms on to complex organisms with brains.

How can we respond to the problem of explaining how consciousness could arise out of a rearrangement of atoms? - consciousness of anything from a simple colour sensation to a feeling of pain or joy. Perhaps this is the biggest 'how' question we can ask. It links our private world of felt experience with our public world of shared ideas, explanations and beliefs. We are steeped in the aftermath of historical answers to this question and to the related question 'why?' Religion, philosophy, science have all attempted to offer their piece of the solution. Can any part be self-sufficient?

Perhaps we can believe that consciousness has an existence entirely separate from the material world inhabited by our chair and this book. We may believe that the soul can exist without any material body. Some religions hold to this general idea. However we may feel that this view has been rendered untenable by increasing knowledge about the brain. Intricate material changes in the brain can be seen with better and better instrumentation. Sensations and feelings consistently accompany these changes in the material state of the brain. If consciousness does not need a material world then why is it so intimately connected to one? Perhaps all this apparent need for material mechanism leads us to an opposing idea. Rather than postulate a materially independent consciousness we simply state that the material world is everything. In current language consciousness is a property which 'emerges' out of the right sort of complexity in material.

Maybe, like me, you find this too hollow. It feels to be a description rather than an explanation. Still unsatisfied we may take some other position. We might feel that what is needed will eventually be found and that it will fit into the content of physical science.

I want to develop a position which doesn't adhere to any of the above. It will pay close attention to the marvelous picture painted by the physical sciences. Crucially it will also pay attention to how the picture gets painted. The fundamental limits to explanation may lie there. There may be much more in the world than science can *in principle* ever see. I don't mean this in the obvious sense of referring to cultural topics such as ethical values or the merit of a piece of art. I mean this at the lower level of 'feeling' the presence of something such as a colour or a sound. This is a level at which it is easier to clarify the issues. However any conclusions can be carried forward and applied to our consciousness of ideas and our emotions.

These questions have been mused upon for centuries. Work in genetic science gives them a fresh cutting edge. The human genome project, genetic screening for disease, genetic determinism and the cloning of

questions from everyday

embryos for replacement cells, all gets intense media coverage. These reports help underpin a notion that our existence seems to stand on nothing more than elaborate molecular mechanism. The possibility of modifying this mechanistic genome using genetic engineering becomes, somewhat ironically, the focus of intensely emotional debate. Where does the experience of emotion reside in our living machine built solely of atoms and molecules from the mechanical blueprint of the genome?

Put this question aside for later and ask another. What needed to happen for a biological world to come into being? One which could produce us? Our conscious life relates to our brain. This brain is an enormously complex structure. As with the rest of our body it makes use of the richest chemistry of any element, the element carbon. This chemistry needs to be rich. It builds life out of the surrounding lifeless material of the planet, generation after generation. However, chemistry does what it does because of the exact way atoms are built from other forces and particles. The smallest change in the strength of these forces would produce huge differences in these structures. The universe could have failed to contain a rich enough chemistry for life of our kind. Life stands on a line of increasing complexity in physical structures.

Why does this arrow of complexity thread its way in tiny regions of the universe? Did the basic forces, first present in the early pre-life universe, contain life as an inevitable outcome? These questions involve a range of informing disciplines. Astronomy, physics, chemistry, biology and brain science. All can be studied in degrees of isolation and some excellent subject-centered books are suggested for further reading. However this book intends to emphasise the layered linking of these sciences. They stack to underpin each other in an order, which maps the time order of the evolution of our universe and within it our planet. The earliest to appear were in the province of physics, the elementary physical forces and particles. They led in time to chemistry, which in its time led to biology. Given more time biology led to brains and so to minds.

If we look at the scientific picture in the round, it is an incredibly beautiful construction in its own right. Seeing the coherence in the picture gives an overwhelming impression of its validity. Scientists build highly successful scientific models of our world. Successful, in the sense that they allow precise prediction and considerable manipulation of the physical world. Of course all the models always have limitations but the biggest problems for their explanatory power probably lie at the extreme ends of our cosmic time-scale, at the origin of the most basic forces and particles of physics and at the manifestation of mind out of brain. Within these extremes the scientific approach to gathering knowledge about the physical world works amazingly well. Single out an issue for study. Observe and then form a theory. Use the theory to predict other effects which should happen if it is correct. Do experiments to bring together conditions which should demonstrate the effects you predicted. If this fails, modify the theory and try again. There are many effects which can distort the cycle. Errors can occur because of genuine error, wishful thinking or plain inability to accept a disagreeable outcome. These are not rare in science. Sometimes two errors cancel, such as when the inability to drop a strong intuition eventually uncovers earlier observational errors. However the approach works because at its root it is evidence driven. It demands evidence which any skilled observer or experimenter can obtain. In the long run its appeal to method is greater than its appeal to authority. In physical science the ultimate test between one idea and another is an *observation* of the world.

The summary above hints at the root of the problem we explore in later chapters. Scientific method requires its subject matter to be observable by others. It deals in what are taken to be 'objective' phenomena which all observers can agree to measure. It has to reduce eventually to quantifiable measures which can then be compared. Consciousness is the unaccounted spectator. Consciousness sits on top of the mound of observations and measurements made by physical scientists but remains unrecorded within the science. Only numbers recording some physical quantity are taken forward to be used later. The practice of physical science does not appear to need any explicit reference to consciousness. We can see the issue simply by considering the most direct sensations we experience. Those we get through our basic senses of sight, hearing, touch, smell and taste. In the sensation we have of seeing a colour, the feeling is not captured by the number which describes the wavelength of a light beam. It is not a term in any of the scientific formulae which describe light. Nor is it a term in the molecular models of the visual brain. How can feelings of joy or love or the sorrow of bereavement be only atomic and molecular configurations? Should we accept that consciousness will remain *in principle* inexplicable in terms of physical science? If this conjecture is true then we must surely

questions from everyday

be very sceptical of grand notions involving 'theories of everything'. If it is true, then perhaps we should be more careful to distinguish between the world which simply 'is' and our partial models of the world. We should avoid mistaking a shadow for the real world.

These awkward questions will become more central as we proceed. Let's start with *(Figure 1.0)*.

This gives a framework for development in the chapters which follow. The vertical column represents the history of the known universe, rising through time from the Big Bang to the arrival of conscious brains. After the Big Bang the early universe cools. Elementary particles and forces separate as the expansion and cooling progress. This results in a universe made of hydrogen, helium, radiation and some basic forces such as gravity. Gravity clumps the gas into high temperature balls which become stars. Cycles of star explosion and formation build the heavier elements including carbon. Many stars capture this heavy material into planetary systems. The Earth is such a planet. Then a separate fundamentally different kind of development occurs. Life begins. Large carbon based molecules interact in ways which lead to a bacteria-like organism. Single celled organisms 'co-operate' to make multicellular organisms. Complexity grows and grows.

The Conscious Universe

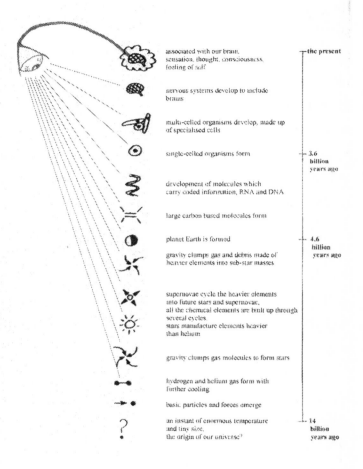

Figure 1.0
Title: **'mind reconstructs its origins'**

The strange loop of mind viewing its origins in inanimate matter. The column of icons charts development from the basic particles and forces of the early universe to our conscious gaze at this process through the eye of science.

Source: own

questions from everyday

Creatures with central nervous systems appear. Brains arrive, which enable sensation and feeling. They further develop to support ideas. Mind has arrived, with profound feelings and values. This is depicted by the human brain and mind at the top of the column in *(Figure 1.0)*. Mind looks outwards. It views the outer world through the channels of the senses. From here it builds pictures and ideas about what is out there. Such an elementary sensation as seeing a white object, doesn't tell us what is really out there. It only reveals the last link of a chain of interactions connecting the something out there to our mind. And if such elementary sensations don't connect us directly with the outer world where are we with much more elaborate ideas which derive from our senses?

Perhaps the most difficult problem lies at the top of the column when mind tries to explain its origin in terms of its own sensations and interpretations.

In the next four chapters we will survey what is said by science about the objects lying in the column which eventually produce the brain. Then in the later chapters we can explore how this physical science picture of the universe fits with consciousness.

CHAPTER 2

the potter's kiln

Let's start our journey by the simple act of looking up to the clear night sky. We sit on a sphere, which has turned to hide the sun from us. The risen moon basks in the sun's light. The light from the moon takes around one and one third seconds to reach us. For the brighter stars, it will have taken their light a few tens of years to get here. Light travels at 300,000 km per second, so one light year (the distance light travels in one year) is about 9.5 million, million kilometres. At jumbo jet speed we would take about 150 thousand million hours to reach one of the brighter stars. That's about 17 million years and astronomically speaking we would still be in our backyard.

We can see one to two thousand stars with our unaided eyes. The dimmest are roughly one hundredth of the brightness of the brightest stars. Stars are fainter because they are smaller, less hot or further away. If we look at two stars of similar size and temperature, one being twice as far away from us as the other, the more distant star will be only one quarter as bright. A star ten times further away will be one hundred times dimmer. This is because the light energy emitted by a star spreads out in all directions. A fixed amount of light has to spread itself over the surface of a bigger and bigger sphere as it flows outwards. The faintest stars seen unaided are a few hundred light years away, up to two million, million jumbo-jet hours.

One of the most obvious constellations in the winter sky is Orion, with the hazy patch in the 'sword', just below the bright three stars of the 'belt'. The hazy patch is a gas cloud in which new stars are forming, 1,500 light years away. That distance is less than one percent of the diameter of the huge disc of stars and dust, the Milky Way galaxy. Our sun is just one of 100 billion stars in this galaxy. We live within this flat disc and see it edge on as a faint strip of stars across the night sky from horizon to horizon. The bright stars, which surround us in the night sky, belong to the nearer part of this disc.

Our unaided vision can even see outside of the Milky Way. If on a clear dark night we look in the direction of the constellation of Andromeda, we can just see a small hazy elliptical patch of light. This is another galaxy, just like our own, only larger. It's one of a cluster of galaxies in our galactic neighbourhood. It's about 2.4 million light years away and more than ten times the diameter of our galaxy. It's 16 hundred times as far as the hazy patch in Orion. This is as far as we can see without a telescope.

Although people have always looked up to a similar view of the night sky, a realisation of the distances involved did not come until the development of astronomical instruments. In fact the knowledge that galaxies existed outside of our own was only properly established in the 1930's.

Galaxies are grouped in clusters. Andromeda is just one member of our local cluster. There are about twenty galaxies of different size and kind within a radius of 3 million light years of the Milky-Way. A typical cluster is 5 million light years across. The distance between clusters is much bigger, typically about 250 million light years. Today, with the most sensitive telescopes, we can see about 12 billion light years. That's about 5 thousand times as far as the hazy patch you can see with your own eyes in Andromeda.

The Conscious Universe

(Figure 2.0) illustrates some relative sizes, from the local street to the observable universe. The absolute distances are too big to visualise. The sketch uses repeated scaling by one thousand times. This scaling can easily be visualised as the one millimetre and one metre markings on a tape measure.

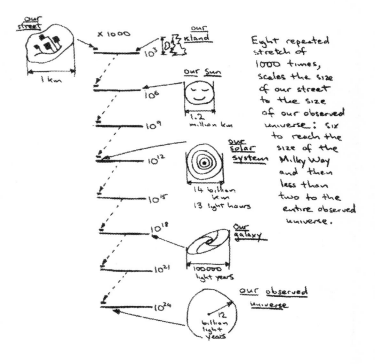

Figure 2.0
Title: **'stretch and repeat'**

Visualise huge distances as just repeated stretches by 1000 times. We can see 1000 times in one view on a tape measure, 1mm and 1 metre.

Source: own

Now that we have a notion of size, let's bring in time. When we look in all directions from us we see a spherical space with us at its centre. As noted above, the outer surface of this sphere, the limit of what we can see with the best telescopes is around 12 billion light years away. It is called the observable horizon. To help us visualise the way space is laid out around us, let us change our spherical space into an onion of the same size! This will help us imagine spherical shells within shells, like a giant Russian doll. Our view from Earth to the Moon is represented by the inner most shell of the onion. This is the closest object to us in the heavens. In the next shell is the Solar System consisting of the other planets and the Sun. Further away are the brighter stars, the Milky Way and other local galaxies. Shells further out contain more and more distant galaxies. As light takes time to travel, the layer which is seen by us as most distant (the outside layer of the onion) is the one which is also seen by us as farthest back in time. This is the layer where typical objects such as galaxies will look the faintest. How far back in time are the farthest seen objects? As it takes their light 12 billion years to reach us, we see events 12 billion years back in time. We see back the same number of years as the distance in light years.

As we look out to each layer of the onion, a whole history of time can be seen by looking at different distances. Of course we must be careful when we take this to be a proper history of the universe. When we look at any particular layer now, we are only seeing one snapshot in its history. We assume that every layer has run a similar history so that the samples we see from different layers can be added together to make a representative 'history' of our observable universe.

A prominent large-scale feature of our universe is its expansion. Let's modify the onion model. Imagine that it grows very quickly. If it grows uniformly, any two points inside will move apart. If you join these two points with a line and mark a third centre point all the three points will move apart as the onion grows. If the distance between the two outer points doubles, the distance between the centre point and each of the outer points will increase by only half this amount. If we time this expansion, then in one-second points which are twice as far away from each other move twice the incremental distance. This means that their velocities relative to each other are twice as great. If we lived on an end point of our line we would see the far end point moving away from us at twice the velocity

the potter's kiln

of the centre point. Because the expansion is uniform the same result would hold looking in any direction. You can illustrate this easily by stretching a rubber band *(Figure 2.1)*.

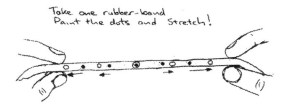

⇄   units of distance moved by each dot in same period of time

•   dot position with little stretching

○   dot position with more stretching

(i) Dots are always equally spaced.

(ii) The farther apart dots are from each other, the bigger their increase in separation after stretching.

Figure 2.1
Title: **'the farther the faster'**

To an observer living anywhere along this rubber-band universe, whose range of view was less than the distance to an end, the same uniform increase in stretch with distance would be seen. Looking to the left or to the right, they would see themselves at the centre of their expanding universe. This would only stop if they could see far enough to include one of the ends. Then the view would show a different maximum speed of stretching in each direction. The observable horizon of our own stretching universe shows general uniformity in all directions.

Source: own

We have pictured our universe as a set of spherical shells like an onion. As we look outwards from the centre in any direction, we automatically look backwards in time. If the universe is expanding with time, the farther away we look, the faster the objects in it are moving away from us. There is an observable effect from this rushing apart. It's called red-shift and it's a property of light waves. Visible light is composed of a spectrum of colours from blue to red. Blue light has a shorter wavelength than red light. When an object sends out light of a certain colour, the observed colour depends upon the relative velocity of the object and observer. One way to visualise this is to watch a stone dropped into a pool. The surface ripples send an expanding set of peaks and troughs out from the centre. If we sail towards the stone then we encounter more of the peaks in say, ten seconds, than if we stayed still. If we sailed away from the stone we would cut through the peaks at a lower rate. The wavelength of red light (the distance between peaks of the wave) is greater than for blue light. If the source and observer move towards each other then the colour observed moves towards the blue (the peaks are arriving faster) and for movement apart the colour moves to the red, the so called red-shift. More generally, this is called the Doppler-effect. It's what you

the potter's kiln

hear when a car or train passes with its horn going. It has a high pitch coming towards you which lowers as it passes and goes away.

Astronomers use spectroscopes to measure the fine spectral lines of starlight. These spectra show that the same chemical elements are present in distant stars. However the frequency of the light emitted by identical chemical substances in distant stars, galaxies and nebulae is shifted towards the red end of the spectrum. The amount of shift is bigger the greater the distance from the source. The universe is expanding. If the measured distance of each galaxy or galactic cluster is graphed against its redshift, then the graph is very close to a straight line. This shows that the speed with which the galaxies are receding is proportional to their distance from us, just as in the growing onion model.

Red shift allows velocity to be estimated. Other techniques are needed to measure distance. Special stars are used. They are well defined types which can be identified as belonging to the distant galaxy under measurement. Their true brightness is known from information on much closer stars of the same type where distances can be measured by geometrical methods. The apparent brightness of these types of stars embedded in a distant galaxy gives a measure of distance. The most recent measurements have used stars entering their supernova phase. The time a supernova takes to flare to its brightest and then fade forever is related to its absolute brightness. Its actual apparent brightness is measured to give its distance. Recent work allows red-shifts to be graphed against distances out to 7 billion light years. If the relationship between red-shift and distance is assumed to continue, unchanged, beyond 7 billion light years, then red-shift alone can be used to estimate the distances of galaxies. It is this indirect measure of distance which places our observable horizon at around 12 billion light years.

There is another effect of the expansion. As we look further and further out, not only do the objects get fainter, because we intercept less and less of the light they emit but also because they are red-shifted. The colour of their light is moved more and more to the longer wavelength end of the spectrum where light has less and less energy. This makes extension of the observable horizon even more difficult than would the effect of distance alone.

the potter's kiln

What is the age of our observable universe? Our onion model helped illustrate the effects of the expansion. It can also be used to estimate the age of our universe. Inside our growing onion we can see that all of the pieces are expanding away from each other. In order to estimate how long the onion has been growing we measure two things. The first is the distance from the centre to a point in the onion. The second is the speed of recession of the same point from the centre. Then we divide the distance by the speed of recession to get the time it took for the onion to expand from a pinpoint to its present observed size. Of course this assumes that the rate of expansion had been constant through time.

In the real universe, measuring how much the expansion rate has changed with time is very difficult. We mentioned above that distance measurements using supernovae have reached out to around 7 billion light years. These can be paired with red-shift estimates of speed to calculate any change in the expansion rate during the past 7 billion years. These results suggest that the expansion is not slowing down with time. The expansion process is opposed by the mutual gravitational attraction of all the mass in the universe. However, as the universe expands the distances between the clusters of galaxies increases and the gravitational forces of attraction decrease. For every doubling of the distances the forces divide by four. Earlier models of the expansion saw it as a race between the total quantity of mass in the universe and its initial rate of expansion. This balance would determine whether it would expand forever or eventually contract. When galactic red-shifts were measured it became apparent that there was no where near enough observed material in the galaxies to account for the steady expansion rate. The visible mass in the galaxies accounted for only one or two percent of the total needed. More recent measurements of the relative motions of galaxies within local clusters have measured this visible mass deficit for individual galaxies. These galactic motions suggest that there is ten to twenty times more unseen mass than shows in the radiating material of a galaxy. This dark matter of the universe may include large dense components such as stray planets, stars too small to ignite (known as brown dwarfs) and neutron stars. It may also include atomic and subatomic particles.

The expansion rate estimates have also changed in recent times. Red-shift measurements have shown that the universe is currently accelerating. A subtle shift from

the potter's kiln

deceleration to acceleration seems to have occurred when the universe was about half its present age. It looks like there may be an additional source of energy involved in accelerating the expansion as the average density of matter falls. This new energy is not presently understood. Ideas which only a few years ago looked ready to settle are again challenged.

The current measured expansion rates can be used to project back in time to suggest an 'origin' time. After taking all the uncertainties into account the consensus for the age of the whole universe is around 14 billion years. However this may be altered by new measurements for the period earlier than 7 billion years. It is always wise to expect uncertainty near the frontier.

Let's turn to the question of what is outside of our observable universe. We need a new model to talk about this. Once we have described this model we will be able to fit the onion model into it. We will be able to see our observable universe inside a bigger universe.

To draw the universe growing in size we need to cheat a little. With three-dimensions needed for space and another for time we will have to drop one of the space dimensions in our sketch and use it for time instead. We

The Conscious Universe

can do this by representing the size of the universe at a given time by a disc instead of a sphere. With this in mind we can look at *(Figure 2.2)*

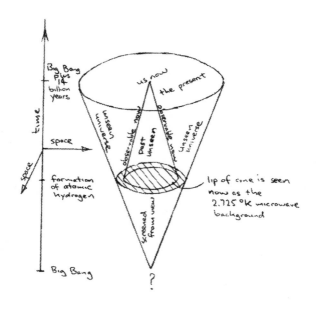

Figure 2.2
Title: 'our observable universe within the larger universe'

    We see very little of the larger universe. By representing the size of the universe at anytime by a disc, expansion can be shown by stacking the discs into a cone. The larger inverted solid cone shows the universe expanding from the Big Bang to the present. Astronomers can only see the surface of the upright smaller cone. Only on this cones surface does the distance of objects from us divided by their time in the past equal the speed of light.

Source: own

the potter's kiln

It shows two cones formed by stacking discs of growing or diminishing size on top of each other. Let's describe them one at a time. We will start with the largest inverted cone which has its point standing on the '?' at the bottom of the sketch. Time runs up the page. Distance runs across and also out of the page towards us. The base of the large cone is the edge-on disc labelled 'the present'. This cone represents the universe emerging from a point (the Big Bang) and uniformly expanding to the size of the top disc representing the present. Notice that we have drawn a simple cone. Its slope is constant over its height. Since time runs up the page and one of the distance axes runs across the page this slope measures speed (distance divided by time). It corresponds to the rate of expansion of the universe. The drawing is for the constant expansion case which is very close to measurement for the past 7 billion year period. If the rate were higher before 7 billion years ago, then the lower discs would get smaller faster. They would converge more quickly towards a point and the universe would be younger. If the expansion was slower before 7 billion years then the lower discs would get smaller more slowly. They would converge more slowly towards a point and the universe would be older. This is a current cause of uncertainty in estimates of the universe's age.

Let's now look at the smaller upright cone in *(Figure 2.2)* which has its point in the top disc labelled 'the present' at the point marked 'us now'. This cone represents our view outwards as previously described using the onion model. The surface of this cone contains everything we can ever see simultaneously 'now'. It is a result of the speed of light. Everything astronomical we can see 'now' is the result of light (or similar radiation) crossing space at a fixed speed. The further away the event the farther in the past it happened. The slope of this cone is fixed by the speed of light and only events on the surface can have the correct ratio of distance divided by time to be there. We can see nothing inside the cone because it is in a past whose light has long passed us by. Above the surface of the cone we can see nothing. There has not been enough time for this light to reach us.

This model shows how small the sample of the universe is which we can ever see: only the slender surface of a cone extracted from within the solid cone of the universe.

As we continue to travel up the time axis we pull the cone of our observable universe behind us. The whole universe continues to expand around us but our observable horizon will stop growing when its speed of recession reaches the speed of light. Then red-shift will have removed all of the energy from radiation. Currently the universe looks set to expand indefinitely. The average density will fall towards zero. The clusters of galaxies will continue to move apart but our local galactic cluster and our own galaxy will follow its own course. Only deep space astronomers will notice the changes.

There is yet another feature inhibiting our view of the early phase of our universe. This is illustrated in *(Figure 2.2)* by the shaded disc at the base of the cone representing our observable universe. This disc is a screen between us and the Big Bang. Before discussing this we need to say a little more about light.

Light fits into a very small part of the spectrum of electromagnetic radiation. Gamma rays lie at the short wavelength (high frequency) end of this spectrum. With lengthening wavelength (declining frequency) there are X-rays, ultra-violet, and then visible light with its rainbow of colours from blue through red. After that there are infra-red or heat-waves, micro-waves and radio-waves. This entire spectrum of radiation can now be observed with modern astronomical instruments. The higher frequency radiations beyond ultra-violet through X-rays and gamma rays are largely stopped by the Earth's atmosphere. The instruments to measure these radiations have had to be placed in space outside of the shielding effect of the atmosphere.

However it is the information obtained from the low frequency microwave end of the radiation spectrum which has been most important to us in understanding the early phase of our universe. Highly directional microwave radio receivers have made measurements of the intensity of this radiation, looking into space in all directions from planet earth. It has been found to be very, very uniform. Out of the plane of our galaxy the radiation density is uniform to within one part in 100,000.

To take the discussion further we need to understand the relationship between temperature and radiation.

If we heat up a piece of metal in a gas flame or a coal fire, the longer we leave it the hotter it will get. Eventually it will approach the temperature of the flame. Even before we can see it visibly glowing, we can feel its infrared radiation on our skin. As it rises in temperature

the potter's kiln

it initially glows dull red, then progressively more orange until eventually if we raise its temperature enough it becomes a bright yellow. The energy peak in the colour or frequency of the radiation is related to the temperature of the emitting material. If we use a hot enough emitter we can take the peak of the emission through blue into ultra-violet radiation. This is the radiation which tans or burns our skin.

We can see another property of radiating material by looking into the centre of a coal fire or a solid fuel barbecue. When the hot coals are all together with their radiating surfaces facing each other, every surface comes into a state of thermal equilibrium with every other. A uniform temperature is established and this temperature determines the colour of the radiation from the centre. When all the radiation is coming from material at the same temperature, all the detailed surface features disappear because all the light has the same colour.

A better example can be seen by looking into the spy hole of a potter's kiln. All the pots which are inside are invisible when the kiln is hot. Their images blend into each other and into the walls of the oven.

The Conscious Universe

Such a radiating system which has reached thermal equilibrium is, rather oddly, called a 'black body'.

Single temperature radiators leave a very distinctive radiation fingerprint. They radiate energy over a range of wavelengths or frequencies in a very exact way *(Figure 2.3)*.

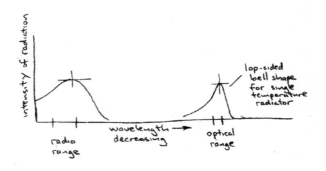

Figure 2.3
Title: **'wavelength yields temperature'**

For a 3000 degree K star surface, the peak wavelength corresponds to visual red. For a 3 degree K radiator the peak wavelength is in the microwave radio spectrum.

Source: own

The peak of the radiation curve uniquely reveals the temperature of the radiator's surface.

the potter's kiln

When the energy of the background microwave radiation surrounding us in space is plotted against wavelength it has precisely the curve described above. It matches this shape to better than 0.03 percent.

To recap, we are surrounded by a uniform microwave radiation with exactly the right shape of energy curve against wavelength to have been emitted by an object in complete equilibrium. This is just like the hot potter's kiln where the material of the pots and the radiation are all in completely balanced exchange of energy.

The peak of the energy against wavelength plot gives the temperature of the radiating source. For the microwave background radiation this is 2.725 degrees K (about minus 270 degrees centigrade).

We have talked about expansion causing red-shift. We have described how this can be used to estimate how long ago the content of our currently observable universe was packed into some small fraction of its presently observed size, around 14 billion years ago. Currently astronomers

The Conscious Universe

believe that the background radiation we are now seeing has a similar age. It has travelled for around 14 billion years across a distance of around 14 billion light years from the outermost shells of our universe. Just like all the other distant light we see it will have been red-shifted. As this light was emitted from the most distant layers it will have the largest recessional velocity and hence the largest red shift. Its frequency would have been very much higher (shorter wavelength) when it was originally emitted around 14 billion years ago. To estimate its frequency at emission we need to know how big our universe was relative to its present size when the radiation was released. Current ideas suggest that the radiation which we now see as this 2.725 degrees K uniform background microwave radiation was released when our observable universe was about 1/1500 of its present size. Since red shift is proportional to distance, the surface temperature of the black-body source required at the time of emission would have been around 1500 times the presently observed value of 2.725 degrees K. That's about 4000 thousand degrees K. That's slightly cooler than the present average surface temperature of our Sun.

the potter's kiln

What caused this uniform emission around 14 billion years ago? At 4000 degrees K atomic hydrogen cannot exist. Above this temperature the future components of hydrogen exist as a plasma of electrons and protons. In this state radiation interacts freely with electrons to achieve equilibrium. If we follow the expansion of our universe forward in time from here, the red-shift due to expansion causes the radiation to be reduced in frequency. As this radiation encounters and interacts with electrons, the electrons receive less energy and therefore re-radiate at a lower temperature.

This process of cooling with expansion continues until it is low enough for electrons to combine with protons to form atomic hydrogen. By 3000 degrees K this process is dominant. The light released during this intermediate phase is the future microwave radiation we observe now after it has been red-shifted to 2.725 degrees K.

The radius of our universe at this early phase would be around 1/1500 of 14 billion light years. That is an object almost as hot as the Sun but only 9 million light years in radius. This is approximately one million times the distance from our Sun to its nearest neighbour stars.

We can now return to *(Figure 2.2)* and see the significance of the shaded disc at the horizon of our observable universe. It represents the 4000 degree K emission at the formation of atomic hydrogen. The thermal equilibrium preceding its release has obliterated any information about the earlier features. It's the potter's kiln, thermal equilibrium obliterates earlier information. Above the shaded disc is the province of observational astronomy. Below it is 'open season' for the theories of high temperature particle physics. This is a lot less direct than observation. It involves lots of inference and uncertain ideas. Don't expect stable consensus.

We can summarise. The amount of red shift (or cooling) in radiation from objects such as stars, gas or high temperature plasmas depends upon how far away such objects are. This is because more distance means a faster speed of recession. We can look at this, another way. If instead of viewing the universe from now, we imagine travelling back earlier and earlier in time, the distances between objects would get less and less. As distances in the past are less the same rate of expansion gives less red shift. Because there is less red shift the wavelength of radiation is shorter which means its energy is higher.

the potter's kiln

We also know that when matter and radiation interact hotter radiation means hotter matter. That's why we need to wear beach slippers or run fast to the sea when we holiday near the equator. To know how much hotter, we need to add another piece of physics. For systems of tightly interacting matter and radiation (so-called 'black bodies' like our potter's kiln or the early dense universe) the energy in radiation is proportional to the temperature of the emitting body raised to the power 4. (That's squared and squared again). For every doubling of the temperature of the body we get sixteen times as much radiant energy.

If we look back in time, for every halving of the size of the universe, red shift halves. As a consequence temperature doubles and so the energy content in the radiation goes up sixteen times. We need some more physics. Energy has mass. This follows from Einstein's famous Theory of Relativity ($E=Mc^2$). Radiation has gravitational attraction just like matter.

When the universe was smaller, the average density of the matter was also higher. The same amount of matter was squeezed into a smaller volume. Half the diameter gives one-eighth of the volume so eight times the density. There are two influences competing. The gravitational effect of radiation is competing with the gravitational effect of matter. From our description above we can see that radiation will win out over matter by sixteen times to eight for every halving in size. Matter will win out over radiation by sixteen to eight for every doubling in size.

The behaviour of the very early phases was radiation dominated. Matter became the major shaping force of the galaxies only after expansion switched the gravitational dominance from radiation to matter. Red-shift denuded radiation's mass advantage.

When? Well this transition is thought to have taken place when our present observable universe was 1/10,000 of its present size, that's about 1.5 million light years in radius or one third of the distance from our Sun to the next nearest star.

In the phases even earlier than this, model building requires a whole bundle of particle physics. To interpret and use these models requires deep specialist knowledge. This knowledge is based on experiments carried out on Earth in particle accelerators. In these experiments matter is taken towards the kinds of extreme temperatures which occurred in the very early-universe.

the potter's kiln

(Figure 2.4) is a sketch of temperature versus size showing some of the major events in the early universe.

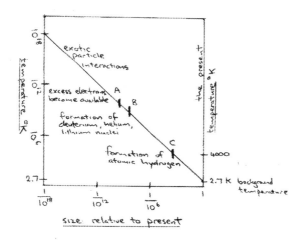

Figure 2.4
Title: 'expanding, cooling, structuring'

        The universe's age, size and temperature and the arrival of atomic material
  When electrons and protons combine to form neutral hydrogen, radiation is freed from continuous interaction with electrically charged particles and can set out on its largely independent journey through space and time. Eventually a very, very minute amount will fall into the telescope of a patient astronomer.

Source: own simplification of fig 5.1, page 139 of 'Our Evolving Universe', Malcolm S. Longair, pub. Cambridge University Press. ISBN 0 521 55091 2 (courtesy of D. Malin and the Anglo-Australian Observatory)

Note the extremes of size. It has expanded from 1million, million, millionth of its present observable size. Note also the temperature extremes, cooling from 1 million, million, million degrees K to a present average background of 2.725 degrees K. The earliest phases of these models are highly speculative. The phases where particles and antiparticles interact to form protons, neutrons and electrons are more stable parts of the model.

This overlap of cosmology, physics and astronomy has attracted a growing number of researchers over the past fifty years. It is still very speculative because the earliest phases of the universe remain beyond the reach of observational astronomy. The extreme temperatures and densities are also outside of the reach of current laboratory physics. New ideas are required to build physical models which work at these extremes. New theories come and go. An open but cautious mind is required.

The distinction between a descriptive or predictive model and an 'explanation' must certainly arise in any attempt to reach an 'origin' event such as the Big Bang. Explanation has to push back to indefinitely higher temperatures and densities. Attempting to cover these extremes, physicists invent new theories. These call up new 'fundamental' entities. Are these entities simply convenient mathematical inventions? This distinction between description and explanation has always been troublesome but never more so than in the physics of the last hundred years. Fundamental physics has acquired something of a split personality. Its theoretical models can predict effects with a high degree of accuracy. But the same models don't reconcile with our most basic intuitions about space, time or cause and effect. This interplay between science's account of the world and our personal perceptions will be developed in the chapters ahead.

CHAPTER 3

down to earth

In chapter 2 we looked out as far as we could see, first with the unaided eye and then with the best instruments available today. This took us to the epoch when our universe was about 1/1500 of its present size. This was when the first atoms came together, atoms of hydrogen. How did hydrogen gas develop into galaxies of stars? How did stars make the larger atoms of the elements such as iron, carbon and silicon, which are incorporated into the construction of planet Earth? Atomic hydrogen formed as the temperature of the early universe fell below 4000 degrees K. The fingerprint of this phase is the 2.725 degrees K microwave background radiation we detect today. This is approaching us very evenly from all directions suggesting a very uniform origin. But how does something which is spread in a very uniform way cluster or clump to become very unevenly spread? How did things change from the very uniform past shown by the microwave background radiation to the very non-uniform present we see in the sky around us?

Think of the situation that results when a large number of objects are uniformly spread throughout a region of space. These objects could be particles such as hydrogen atoms or grains of dust or much larger pieces of material. They would all behave in a similar way. As these particles have mass, each particle will be attracted to every other one through the force of gravity. If everything is uniform there is no net force to move any particle in any particular direction. Everything remains uniform. Introducing the smallest disturbance into this uniformity will cause an instability which will develop by its own process. This is because the force of gravity increases the closer two objects approach each other. It multiplies by four for every halving of the distance. When some disturbance occurs to produce a small region with an increased number of particles, these particles will attract each other more. This is because they are closer to each other than to those beyond the edges of the group. They move towards each other and form a cluster. This cluster now has a collectively stronger gravitational pull to capture other particles that come close to the outside of the group. This is because a passing particle can be closer, simultaneously, to more particles in the group than in the surrounding average background. As its mass grows, the cluster sweeps up more particles, becomes more massive and gets a stronger gravitational attraction.

Now where did the initial irregularities come from that led to multiple clusters of particles?

The answer is probably that there were many tiny traces of non-uniformity within the hydrogen cloud each of which seeded the start of later clustering. The early anomalies in the hydrogen distribution will probably have come from some earlier non-evenness in an even earlier phase. Gravitational attraction starting from the seed of some non-uniform density in the hydrogen cloud pulls more hydrogen atoms together. The more the process runs the more mass is in a given volume and so the stronger is the gravitational pull. So atoms pulled from outside of the clump arrive at higher and higher speeds. When the density is high enough a significant number collide with other atoms already in the clump, transferring momentum and raising the speed of these atoms.

As the mass of the cloud grows the accelerating atoms falling into the cloud gain more and more kinetic energy. Through interaction the atoms share out this energy by raising the average speed of all the atoms in the clump. Now a measure of the speed of atoms or molecules moving around in a gas is better known as the temperature of the gas. As the gas clump forms and grows energy gained from the effect of gravitational attraction raises the temperature of the gas. This process would continue indefinitely but for the emergence of an opposing effect. As the temperature rises the outward force experienced by incoming molecules during collisions with fast outgoing molecules generates an outward pressure. This is in the opposite direction to the gravitational attraction of the clump. The higher the speed of the molecules inside the clump, the more they can escape the gravitational bond of the clump and leave it behind.

On a more general note, this is an example of a very important feature of complex interactions, 'dynamic equilibrium'. The growing effect of one process generates conditions which strengthen the hand of a second process which acts to counter the first. In the real world the more we look at large systems the more processes we see intertwined in this state of dynamic interaction. Depending on the complexity of the interactions the overall outcome may behave smoothly or erratically.

As the mass of the gas cloud grows, the temperature required to hold back gravitational pressure also grows. The temperature will thus grow until it is sufficient to stop it growing any more. The temperature reached depends on the mass of gas collected. This is because the energy brought into the clump was the kinetic energy of this mass falling inwards. If there is sufficient mass in the clump the temperature will exceed the ionisation temperature of hydrogen and the clump will move into the plasma state unbinding the protons from the electrons. This is just the opposite of the cooling process which formed atomic hydrogen in the 4000 to 3000 degrees K cooling phase of the earlier universe.

If clump formation is happening all around within the hydrogen gas, eventually most of the hydrogen will have gone to a clump. Further gains from the kinetic energy of accretion will not be possible. Since the hot clump will always be losing energy by radiation it will cool down in the absence of any other internal source of energy. This is the state of the so called brown-dwarf. A brown-dwarf is a star which did not start with enough mass to raise its temperature to the critical point where another internal source of energy becomes available. This is the energy of nuclear fusion. Stars which start with a mass of less than 1/15 of the mass of our Sun are thought to become brown-dwarfs. After gathering all the mass they can they begin to cool, radiating at longer and longer wavelengths into the infrared.

At masses higher than this the internal temperature of the star rises through the gravitational processes we have outlined to the point where the protons in the plasma can combine to form helium nuclei. The mass of the helium nuclei is less than the separate mass of the four protons joining to make it. This mass difference appears in an equivalent amount of energy. The conversion is described by Einstein's equation ('e' equals 'm' 'c' squared). Nuclear sources of energy are unimaginably larger than everyday chemical reactions such as burning gasoline. If one gram of mass could be converted to energy every second then 90 million megawatts of power would be released. About

1/120th of the mass of every proton joining to make a helium nucleus, is converted into energy.

The temperature of the fusion process is enormous. The Sun has a core temperature of around 16 million degrees K. It is held together by the gravitational pull of all its parts. It is pushed apart by thermal pressure generated turning hydrogen nuclei into helium nuclei. These processes are in dynamic equilibrium. The process of building helium from hydrogen is only the start of a chain of high temperature interactions which happen inside stars. At temperatures of tens of millions of degrees further combinations of protons and neutrons from different nuclei create completely new nuclei. The configurations of protons and neutrons which stay together to form stable nuclei is governed by the energy involved in holding them together. Nuclei range from the very stable to the highly unstable. The most unstable have large numbers of particles and so are heavy elements. Unstable nuclei break up again into parts, giving out energy. These are the radioactive elements such as uranium. In general light nuclei fuse (fusion) and heavy nuclei disintegrate (fission).

There are many stages in the process of building the elements from hydrogen. The stages are spread through several generations of stars. Some first generation stars explode and supply their output of lighter elements to be swept up with more hydrogen in further clumping. The whole high temperature process repeats. The more massive the second-generation star, the higher the core temperature reached and the heavier the element manufactured. In stars with masses more than fifty times the mass of our Sun, after using up all the available hydrogen, their centres collapse again under gravitational pressure. Temperatures rise even higher and further phases of nuclear fusion occur to build carbon, oxygen, iron and upward.

Each cycle causes variations in the core temperatures and energy output of the star. Across a range of stars at different stages in their life history this is observable as variations in their colour with brightness. These stars may then shed material as instabilities occur in their balance between gravity and thermal pressure as they exhaust all available nuclear fusion processes. They run out of "fuel" in their core, collapse internally and blast off their outer shells.

Eventually the nuclei needed for all the naturally occurring elements are built. Through further explosive shedding of outer shells from stars this rich cocktail is made available for further gravitational clumping. This makes possible the eventual formation of element rich planets such as Earth, including its occupants. The carbon atoms, calcium atoms, oxygen atoms and all the others which make up the bones, soft tissue and blood of our bodies, had their nuclei made inside a star.

Since we've tried to summarise the birth of stars, we should spend a few minutes on their death. As with almost all other things about stars their mass is the determining factor. There are three outcomes. For stars about the size of our Sun, the likely outcome is instability as hydrogen is exhausted. An outer shell will blow off into space and the shrinking core will temporarily rise in temperature but not enough to ignite the heavier fusion processes. It will form a very bright small star called a white dwarf, perhaps $1/100^{th}$ of its previous diameter. With no mechanism to sustain its internal energy generation, it will cool down through radiation loss until it becomes a cold invisible object.

And what happens to stars more massive than our Sun? This depends on how massive they are. Next up in mass to the white dwarf is the neutron star. In the neutron star the gravitational pressure at their core is great enough to cause protons and electrons to combine to form neutrons. The neutrons have no electrical charge and so do not repel each other. They pack very closely together into extreme densities. At these densities our Sun would be less than 1/50,000 of its present diameter. The period of the collapse of a star to form a neutron star is the source of the short-lived burst of light which we call a supernova.

For a star more massive than a neutron star the end-state is even denser. This is the black hole. The forces which keep the neutron star from collapsing further cannot hold back gravitational pressure. The object continues to shrink into a smaller and smaller volume. Gravity is so intense that radiation or matter falling directly into the object cannot overcome the gravitational force to re-emerge. However, whenever gravitational collapse occurs, there is almost always some rotational motion present. The momentum in this rotation is preserved as the object shrinks. Matter which falls towards the black hole spirals at an increasing rate (just like water flowing out of a bath or sink). It rises in temperature as kinetic energy converts into heat. Once the material is hot enough to ionise, the spiralling charged particles start radiating energy. A basic property of electric charge is that it radiates when forced to change speed or direction. Black holes as accelerators of electrical charge are even more efficient at turning mass into energy than the nuclear fusion process inside stars.

How dense does an object have to be to end-up as a black hole? Incredibly dense! To initiate further gravitational collapse to a black hole the mass of the Earth would have to be compressed into a diameter of less than 2 centimetres.

The production of energy from black holes has been seen on a vastly grander scale than we have just described. There are galaxies where a relatively small central object is observed which can outshine the entire light output of all the stars in the entire galaxy. These objects are called quasi-stellar object or quasars. These energy generators are believed to be massive black holes surrounded by huge amounts of hot ionised material spiralling inwards under intense gravitational pull. They are the largest singular energy sources within galaxies. In one observed galaxy, the output from the central quasar is 1000 times greater than the collective output of all the galaxy's stars.

Rotational collapse of charged matter into black holes is also held to be the source of recently observed intense gamma ray bursts. These are thought to be emissions from the early epoch of gravitational clumping after the formation of hydrogen.

Observational astronomy shows that stars are arranged in huge groups, the galaxies. Our Milky Way has around 100 thousand million stars. The dynamics of its internal motions governed by gravity suggest it must have a total mass of at least ten times that of its radiating stars. The overall distribution of the galaxies is uniform on a large enough scale but looked at in smaller regions the galaxies cluster with large irregular voids between. A good analogy is a sponge. The bigger the pieces the more they look alike: the smaller the pieces the greater the differences.

down to earth

(Figures 3.1, 3.2, 3.3 and 3.4) shows a range of images obtained in the last few years, by ground and space based telescopes.

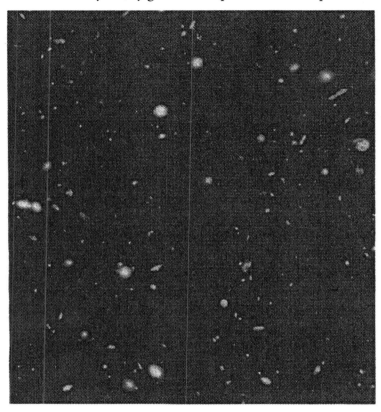

Figure 3.1
Title: **'Hubble Deep Field'**

A tiny region of sky, covering less than 1% of the moon's diameter, taken from the Hubble telescope in orbit around Earth. The fainter galaxies in this field are thought to be 10 to 12 billion light years away. So they are seen as they were 10 to 12 billion years ago when they were relatively young and perhaps only one or two billion years after the first formation of hydrogen in the very early universe. Gravitational clumping of this early gas formed the galaxies and created the stars.

Source: Scanned from page 163 of 'Visions of Heaven', Tom Wilkie & Mark Rosselli, pub. Hodder & Stoughton, ISBN 0 340 71734 3. (Credit: R. Windhorst, Arizona State University and NASA)

The Conscious Universe

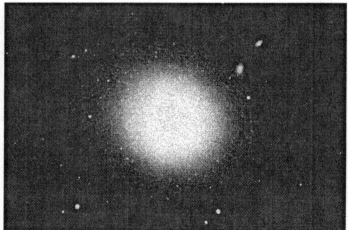

Figure 3.2
Title: **'Galaxies'**

An elliptical galaxy and a spiral galaxy. The reddish tinge closer to the centres of these galaxies comes from older star populations whilst the bluish colour of the outer parts comes from younger stars.

Source: Scanned from fig 4.1, page 99 of 'Our Evolving Universe', Malcolm S. Longair, pub. Cambridge University Press. ISBN 0 521 55091 2 (courtesy of D Malin and the Anglo-Australian Observatory)

down to earth

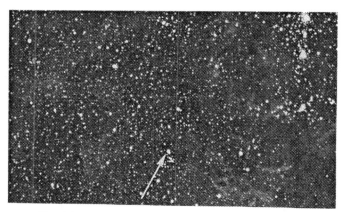

Figure 3.3
Title: 'supernova'

The same area of sky before and after the explosion of the star that became SN1987A. The event was seen from Earth on 24th February 1987. Its debris, together with more gas and dust may eventually be swept up by gravity to make another star or planet, depending on the mass accumulated. Through a series of such star events, the whole series of the heavier chemical elements is built up from the lightest element, hydrogen.

Source: Scanned from fig. 3.9, page 77 of 'Our Evolving Universe', Malcolm S Longair, pub. Cambridge University Press. ISBN 0 521 55091 2 (courtesy of D Malin and the Anglo-Australian Observatory)

The Conscious Universe

Figure 3.4
Title. *'earth and moon'*

Apollo 8 astronaut photograph. Two objects formed by gravity about 4.5 billion years ago but lacking the mass to ignite the nuclear fusion processes of a bright star.

Source: Earth 2000 Ltd. England, 1998. Photo source NASA.

down to earth

Let's move closer to home and turn to the origin of planetary systems such as our solar system. A more detailed study of the formation of stars shows there are other effects to add to those of gravity, kinetic heating and nuclear fusion. Complications occur if there is any rotation in the collapsing clump. Just as when a rotating ice skater pulls in his arms, the rotating gas speeds up as its mass moves into a smaller space. These forces add to the thermal pressure, resisting further collapse in the plane of rotation. When the gas is hot enough to have free electrons, their rotating motion generates interacting magnetic fields within the clump. So there are new forces which cause new behaviour.

In second generation stars there is dust made of heavier elements which changes the way heat is radiated out, cooling the star and allowing it to pull in even more mass. In more recent work these effects have been linked directly to astronomical observations. Huge jets of molecular gas are seen being ejected from the poles of newly forming stars. Flattened discs of gas and dust are seen stretching far out around the equators of rotating stars. Within these discs of gas and dust further gravitational clumping occurs. This may be the process which forms planetary systems in orbit around a star. Observation of stars other than the Sun (with space based telescopes like Hubble) is giving insights into the birth of planets.

Once we have the planet Earth clumping out of a disc of dust and gas the age-old cycle begins. Gravitational collapse converts kinetic energy into heat. Outward thermal pressure resists further collapse. As suitable temperatures are reached, the elements such as silicon oxygen and iron combine through chemical interaction. Heat generating and heat loss mechanisms cause internal convection of materials hot enough to flow. From these physical processes came the rigid silicate rock lithosphere or crust, the convecting flowing mantel of silicate rock and the molten iron core.....pre-biological planet Earth.

And when did it form? Evidence from the decay of radioactive elements suggests about 4,500 million years ago.

The planet we now see has been greatly modified by the emergence of life. The earliest micro-organisms interacted with the planet's surface to produce new conditions. These new conditions supported the emergence of more elaborate forms of life. Such interactions, back and forth, have built and continue to build, the surface and

down to earth

atmosphere of the Earth. We shall look at the arrival of life in the next chapter.

In the previous two chapters we have painted a picture of the development of our universe from an early phase of relatively tiny size and enormous temperature. If we ask what underlying forces shaped this development, then the current views of physicists in this field would invoke four forces. Four are required to build a model with enough richness to deliver good levels of prediction which test successfully against observation and experiment. There is theoretical activity within physics, which tries to show that the four forces originate from a common root. This linking only shows at the highest temperatures of the early universe. The four forces develop independence as temperatures fall. Earliest means less than ten million, millionths of a second after the Big Bang and hotter than one thousand, million, million degrees K. Even there, only two have been theoretically related. It is suggested that a further theoretical unification would only be applicable at earlier and earlier epochs when temperatures were hundreds of times higher.

For our story we will not venture so hot or so early. Physicists believe that the four forces began to act as essentially separate forces well before the first hydrogen atoms appeared in the universe.

These four forces are the force of gravity, the electromagnetic force, the strong force and lastly the weak force.

Let's draw a thumb nail sketch of where these forces are operating in our world.

First gravity…..

Gravity is a universal force of mutual attraction acting between all objects with mass. It is completely independent of the type of material.

It shapes the large-scale universe by bringing material together in several levels of clumping, from gas clouds through stars to galaxies and clusters of galaxies. It creates the conditions in which nuclear energy generation can occur within stars. It also creates the conditions and supplies the energy for the biggest radiation sources such as the quasars. It also made Newton's apple fall to the ground.

The theories or models of General Relativity link gravity to time as well as space. Where gravity is very strong there is a slowing down of all physical processes.

The second force is the electromagnetic force. This is the force that binds the electrons around the nucleus in atoms and binds atoms together into molecules and molecules into everyday objects. It is the force in chemical bonds. Electrons jumping from higher to lower energy shells in their atoms involve the electromagnetic force in emitting light. The same process through a bigger energy difference generates ultra-violet rays, X-rays and gamma-rays. Less energetic electromagnetic processes produce the radio waves carrying television and radio programmes. The branch of physics which studies the connection between matter, electromagnetic force and radiation is called Quantum Electro Dynamics or QED for short. It underpins the contemporary understanding of chemistry.

The third force is the strong force. This is the force that holds the parts of the nucleus of the atom together, the protons and the neutrons. It is the force involved when splitting or fusing the atomic nucleus. It powers the stars when gravity brings enough matter together.

The fourth force is the weak force. It is the force which causes the decay of one particle type into another, such as free a neutron into a proton. When neutrons are not free but are inside a nucleus the effect of the force is altered by the binding energy of the strong force. The effect of the weak force is to test the stability of any particular nuclear configuration. Those which fail disintegrate and throw off particles or radiation. They are radioactive.

(Figure 3.5) is a table which summarises the relative strength and range of these forces. It shows a world governed in realms defined by size.

You can see how the strong and the weak forces drop out from having any direct effects on the motions of objects further apart than the size of an atomic nucleus because their range is so small.

In contrast, both the electromagnetic and gravitational forces are long distance forces. They decline with the square of distance. Although the electromagnetic force is vastly greater in strength than gravity it only acts between electrically charged particles. So the relative weakling, gravity, is left supreme in determining the motions of material on a scale large enough to be electrically neutral.

down to earth

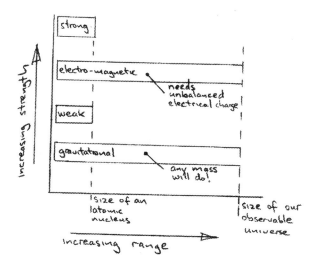

Figure 3.5
Title: 'four forces table'

Gravity rules in governing the motions of the large-scale objects in the universe even though it is the weakest force. The gravitational force of attraction between two protons is 10 to the power 36 less than the electromagnetic force of repulsion caused by their positive charge. However for distant atoms with cancelling positive and negative electrical charges there is no net electrical force.

Source: own

We experience two of the four forces directly in our everyday life. We lift objects against gravity. We feel the electromagnetic force as we tear a piece of paper in two. But none of us can pull the nucleus of an atom apart against the strong force.

The Conscious Universe

These basic forces account for a world made up of minutely small strong structures, the nuclei, encased by much weaker structures of electrons. And finally collections of these atomic structures such as ourselves, are bound by an even weaker force to the surface of our planet. Only the huge mass of our planet gives the significant effect which is our weight.

The current 'standard model' assumes that the present universe came about through the relentless action of the four forces outlined above. The 'standard model' develops and changes with time as new observations and experiments fail to fit into previous models. Sometimes the changes are evolutionary and sometimes revolutionary. There are many examples of revolutionary changes in science, from biology to physics. The replacement of Newton's Theory of Gravity by Einstein's is often quoted. Einstein's theory could account for subtle changes in the orbit of Mercury around the Sun, which Newton's failed to do. Another example from the early 20th century is the quantum theory of light. This embodied ideas from Plank and again Einstein. The central problem in this case was the inability of the existing theory to account for the bell shape in the spectrum of black-body radiation. In chapter 1, mention was made of recent observations which suggest that the expansion of the universe is speeding up. These new observations have caused conjecture that a fifth fundamental force may have to be added to the model.

There are countless examples of new theories for old in the progress of science. The important point is that history suggests that it is very unlikely we have come to any final position. Scientific theories and models are just that. They are not natural laws. *They are human made.*

In this chapter we have outlined the story of how the forces of physics led to the chemical elements and how these elements came together to form planets such as the Earth. We will now move on to look at how the chemical organisation of matter led to life.

CHAPTER 4

climbing up complexity

Astronomy and physics tell us that a set of basic ingredients emerged in the early universe. A group of particle-like objects, protons, neutrons, electrons and a group of force-like objects, gravity, electromagnetic, strong and weak. We have followed the consequences of this to the production of pre-biological Earth. We now want to look at the arrival of life. How did the same fundamental set of interacting particles and forces as developed the planet, introduce life?

Let's proceed step by step. Life needs chemistry. Chemistry needs atoms. The basic protons, neutrons and electrons which make up the atoms of the chemical elements, were produced in the oven of the Big Bang. However, only the lighter elements such as hydrogen and helium were on the first chemical menu. It required many cycles of re-heating, inside stars, before a menu of heavier elements developed. Only when chemistry had a rich enough repertoire could life appear.

Every chemical element has a different number of electrons arranged in shells around its nucleus, controlling its external chemical reactions. Only the electrons in the outermost shell interact with the outermost electrons of other atoms, binding the atoms together to form the molecules which make the materials of our everyday world. Two atoms of hydrogen join to one of oxygen to make water. Two carbon atoms, six hydrogen atoms and one oxygen atom can join up to make alcohol. The arrangement of electrons on the surface of the carbon atom gives it an especially rich range of chemical interactions. Because of this carbon based molecules are central to the chemistry of life.

Chemistry is all about the three-dimensional fitting of atoms together. To get an electrically balanced molecule, negatively charged atoms will 'donate' extra outer electrons to positively charged atoms which are short of outer electrons. This is called ionic binding.

Other chemical bonds occur where electrons are shared among electrically neutral atoms. These are called covalent bonds. Atoms project their bonds in different directions depending on their particular electron configuration. The different atoms are driven by electrical forces to align their bonds in space. This gives molecules their 3-dimensional shape. A third type of bonding occurs when the shape of the final shared electron cloud around the electrically neutral molecule is polarised, that is slightly more positive at some parts, slightly more negative at others. Around hydrogen covalently bonded to another atom there is a slightly positive charge. Around covalently bonded oxygen there is a slightly negative charge. As opposite charges attract, hydrogen and oxygen containing groups form weak bonds known as hydrogen bonds. This is the force of cohesion which makes water liquid.

Although covalent bonds may be strong along an atom to atom axis, they can in some cases rotate freely around this axis. This allows the construction of strongly connected, but floppy structures. A large molecule may be able to fold on top of itself by rotating some bonds. The preferred folding arrangement will depend on the weaker forces such as hydrogen bonding described above. These more subtle secondary forces result in the huge range of shapes in protein structures which are fundamental to living tissue

Because of all these numerous weak forces a large molecule may be stable in a range of alternative shapes. One shape may be harmless and one may be disastrous, as in the case of the prion protein found in the brain tissue of variant CJD sufferers (Figure4.0).

The shape of electron clouds around atoms controls the shape of molecules from the simple linking together of atoms to the elaborate folding and twisting of the molecules of living tissue. The whole of chemistry is a 3-dimensional jigsaw, with the chemical elements as the basic pieces.

Now a particularly simple case of shape in chemistry is seen in a crystal. Here the regular geometry of the atoms can act as a template for the crystal to grow while maintaining its shape. The template of atoms on the exposed surface orders the arrangement in which additional atoms join. If a crystal of salt is put in a concentrated solution of sodium chloride, sodium and chlorine atoms attach to the existing lattice ofatoms making the crystal grow.

climbing up complexity

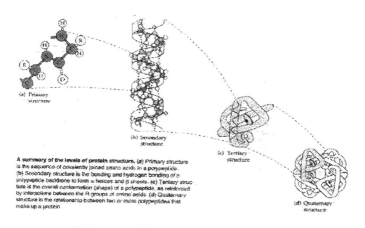

(a) Primary structure
(b) Secondary structure
(c) Tertiary structure
(d) Quaternary structure

A summary of the levels of protein structure. (a) Primary structure is the sequence of covalently joined amino acids in a polypeptide. (b) Secondary structure is the bonding and hydrogen bonding of a polypeptide backbone to form α helices and β sheets. (c) Tertiary structure is the overall conformation (shape) of a polypeptide, as reinforced by interactions between the R groups of amino acids. (d) Quaternary structure is the relationship between two or more polypeptides that make up a protein.

The tertiary structure of the prion proteins involved in mad cow disease

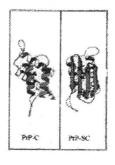

PrP-C   PrP-SC

PrP-C is the normal cell marker protein. PrP-SC is the "mutant" form which builds up to produce the amyloid plaques in CJD.

Figure 4.0
Title: 'connecting, sheeting, twisting, folding and fitting.....the geometry of of proteins'

Source: Scanned from Figure 3.27, page 82, 'Biology', third edition, Neil Campbell, pub. The Benjamin/Cummings Publishing Company, Inc., 1993, ISBN 0 8053 1880 1 ... combined with private source for lower part (Frank Prior).

Crystals often have quite weak bonding in one plane or another. Small pieces of fractured crystal can seed the growth of new crystals from an appropriate supply of atoms. The supply atoms must have the mobility to reach the correct places on the seed crystal to bond. This happens if the crystal is in a concentrated solution.

There is a similarity here to the chemistry in simple living organisms. They too are built up by assembling molecules from the surrounding environment, fitting them together in an ordered way using a template. In both cases there is a chain of replication, with the template dictating the order in which chemical structures are joined together.

climbing up complexity

Let's progress to a molecular structure which is more fundamental to life. It has one of the most commonplace acronyms around today, DNA, short for Deoxyribo Nucleic Acid *(Figure 4.1, 4.2)*.

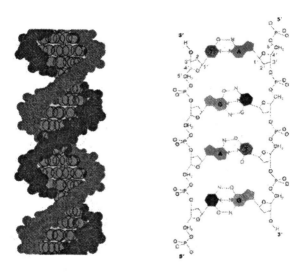

Figure 4.1
Title: **'The structure of DNA'**

The DNA double helix, unrolled to show the ladder structure of sugar-phosphate backbones (the uprights) and cross linking base-pairs (the rungs). There are only four base types whose ordering forms the basis of the genetic message

Source: scanned from figures 11.5 and 11.9, pages 318 and 320, 'An introduction to Genetic Analysis', Anthony J. Griffiths et al, pub. W.H. Freeman and Company, New York, 6th edition 1996, ISBN 0 7167 2604 1

The Conscious Universe

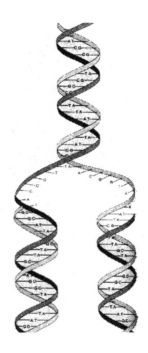

Figure 4.2
Title: **'replication of DNA'**

A DNA molecule unzips from an end to expose two complementary strands. Free bases of the required matching type, A,C,T, or G, assemble themselves along the two unfolded strands to make complete replicas of the original molecule. So one DNA molecule becomes two and the coded recipe expressed in the order of the bases is available to be passed on.

Source: scanned from figure 11.12, page 322, 'An introduction to Genetic Analysis', Anthony J. Griffiths et al, pub. W.H. Freeman and Company, New York, 6th edition 1996, ISBN 0 7167 2604 1

climbing up complexity

Covalent and hydrogen bonds are present, just as in the protein example in *(Figure 4.0)*. The individual shapes of the proteins and of DNA result from subtle manifestations of the basic electromagnetic force. Salt and DNA, although very different also owe their shape to the same fundamental physical force.

DNA is the chemical basis of the genes. It is formed from a template like a crystal but more significantly, it carries plans coded in its molecular structure. As with crystal templates, a suitable environment is needed for action. A suitably matched chemical environment enacts the DNA plans. In such an environment these plans carry the information needed to manufacture proteins. More astoundingly, the DNA plans also contain all the information needed to organise these proteins into a complete organism.

The information required to control all these higher level events is contained in the order sequence of four molecules which make the basic building blocks of DNA. These four molecules are, adenine, guanine, cystosine and thymine, usually abbreviated to A,G,C,and T. *(Figure 4.1)* These four symbols are the alphabet of life. Their sequence gives the instructions for the building and maintenance of everything in the body. For example in protein manufacture the order of the bases is 'read' in

The Conscious Universe

blocks of three, each block coding for one amino acid. These amino acids are then joined together to make new proteins. So the 'words' of the genetic code are 3 letters long from an alphabet of 4 letters.

To hold all the information needed to build a human, 46 separate molecules of DNA are required. This is about 3 thousand million bases, equivalent to the number of letters in 1 million pages of written text. This set of 46 molecules is called the human nuclear genome. *(Figure 4.3)*.

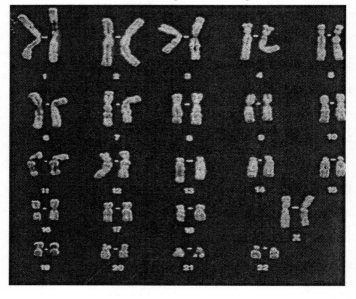

Figure 4.3
Title: 'the human genome'

The human genome is a set of 46 chromosomes or DNA molecules. The set consists of 23 pairs of similar chromosomes. One pair comes from the father and the other from the mother. Chromosomes can only be seen as shown in this optical microscopic image when they are in condensed form after copying and before cell division. At this stage there are two complete copies of the 46-chromosome set.

Source: page 10, 'The Daily Telegraph', Thursday December 2, 1999

climbing up complexity

Twenty-three come from each parent. Copies of these pairs are kept in every cell of the body except for some specialist cells producing sperms. These have only a single set of 23 chromosomes.

The deeper we look, the more we see ingenious mechanism. For example, look at how the recipe for a human being is passed on through sexual reproduction. How is a set of 23 chromosomes extracted from each of the sets of 46 present in the mother and father's body cells? These sets of 46 are organised as 23 paired-up paternal and maternal DNA molecules. Each molecule of the pair has genes for the same functions in identical positions along its length. The process begins inside specialist cells in the testes or ovaries. The first stage breaks the paired-up paternal and maternal chromosomes at two or three identical places. These pieces are then crossed over to make every chromosome have a mixture of pieces from paternal and maternal sources. This is followed by two stages of cell division which pull apart random sets of 23 chromosomes from the crossed-over set. In the case of sperm this process produces four distinct sets of 23 chromosomes. Each gets a head and a tail and becomes a functional sperm. In the case of eggs the two stages of cell division also produce four distinct sets of

23 chromosomes but the egg does not complete the second division until after it is fertilised by a sperm. At fertilisation the 23 sperm chromosomes are kept separate from the 46 chromosomes still in the egg. Only at first division of the fertilised egg to form two cells does a set of 23 get selected to combine with the sperm's set of 23. The extra unused egg chromosomes are ejected from the cell cluster during the two stages of division. The new paired up set again holds paternal and maternal DNA separately in each molecule of the joined pair. In all the subsequent cell divisions building the tissue of the embryo this 46-chromosome set is replicated and a copy kept in each cell nucleus. During development of the embryo selected body cells migrate to the future sites of testes and ovaries. Even before birth the whole cycle is being set up for the next again generation.

The first few divisions of the cells of the embryo are timed and controlled by the outer cell or cytoplasm. The interplay between DNA and the cytoplasm is subtle. Cloned sheep have been developed from identical nuclear DNA inserted into different cytoplasm. These sheep show differences in appearance and behaviour. At present it is not known whether the differences in the cytoplasm have had an effect on the expression of the nuclear DNA.

Above we described how initial differences in the DNA of mating partners is increased very rapidly by the kind of mixing which occurs in sexual reproduction. But where did the differences in DNA come from in the first place? There are additional background processes which are steadily feeding changes into the molecular structure of the DNA. Errors occur during the copying process at cell division. Spontaneous changes can occur when the cell is not dividing. Molecules of DNA can break and rejoin in a different order. Pieces can break away from one chromosome and fit into another. Segments can turn around and become inverted. These changes are called mutations. There are mechanisms of chemical 'proof reading' within the cell nucleus to try to repair these 'errors'. These mechanisms spot inadmissible coding and insert molecules to try to repair the damage. After all of this there is a low-rate background error rate. These mutations can occur in any type of cell. There are two cases to consider. The first is when the mutation occurs in the DNA of body cells which are not connected with the production of sperm or eggs. Here there is no transfer of the mutation to the next generation. When cells of the body reproduce they replicate exact copies of their

DNA. The process of cell division puts identical cells side by side. So a genetic error in one body cell goes on to produce a cluster of defective cells. There are mechanisms to control abnormal cells but if the genetic abnormality involves these control mechanisms then cancer may be the outcome.

The second case to consider is when mutation occurs in the DNA of cells producing sperm or eggs. In this case there is transfer to the next generation in two senses. First, the mutation will express physically in the offspring. Second, the mutation will be passed in the sperm or eggs of the offspring to subsequent generations. The outcome of mutation in the DNA of the reproductive cells is to produce a range of possible changes in the next and following generations.

As noted above, body cell DNA mutations are not passed to next generations. However, if such mutations reduce parental ability to nurture then the viability of the next generation may still be effected. The transferable mutations are taken to be the ultimate source of variation in the theory of evolution. Some mutations allow living organisms to survive and thrive better in their environment, increasing their rate of replication. Those which have the DNA sequence which best suits their

climbing up complexity

immediate environment do best. Those with sequences less well suited may die young without passing on their DNA, or may not mate as often, or may not nurture their offspring to reproductive age. These changes in DNA and the natural selection of the environment are the two mechanisms which the theory uses to account for the phenomenal growth in the complexity of biological objects, from micro-organisms, plants and animals, to us.

How the genes control features is key. A random mapping between genes and their physical effects would produce different possibilities compared with a coherent mapping where for example a single gene influenced the length of a limb. Because of this, coding structures will themselves have been subject to natural selection. Some coding schemes would make more effective use of mutation and natural selection and so give better access to adaptation.

Let's return to the crystal analogy again. The connections and positions of every atom in both salt crystal and strand of DNA are fixed by the same basic electromagnetic forces. At this level of inspection, 'what you see is what there is', nothing more. Both the crystal and the DNA act as a template. However DNA also operates on another level, a level which is invisible at the first level of inspection.

DNA codes for something on another structural level. It is a plan, a blueprint, a recipe. You get more than you see! It's rather like the two levels of a musician's score. At one level it's only marks on a piece of paper.

At another it represents music. But it only becomes music when the marks are interpreted through the action of playing. It is the same with DNA. It needs to have suitable chemical surroundings to action its message. It plays into the chemical machinery of proteins. It needs a matching environment for its activities. This is provided by the cytoplasm of the living cell.

The coding of information inside the DNA molecule is the breakaway point for the rising level of complexity found in living organisms. We have already discussed the seeding of crystals from pieces of previous crystals and considered it to be a mechanical template. Rows of atoms with available bonds fit with complementary atomic partners from the surrounding environment, a self-template. Imagine a more elaborate kind of three-dimensional template which projects bonds to match different types of molecules. A template able to pick up different free molecules with the correct alignment to join and make a new larger molecule. Such three-dimensional templates are found everywhere in living organisms. This is how the enzymes work. Different enzymes organise more basic molecular components together and zip them into proteins. As a result of secondary forces (hydrogen bonding and other similarly subtle manifestations of electrical force) protein molecules twist, fold and roll into a three-dimensional shape *(Figure 4.0)*.

This progression can be explained stepwise up to a certain point. Let us call the growth of crystals, a very simple type of template coding, level-1.

One step higher, we have one substance 'coding' the manufacture of a different substance. Let us call this process which we see in enzymes, level-2.

The next step is where coded information results not just in a particular set of materials being produced but in these materials being arranged in space to form a complicated object, such as a hand or a brain. Let us call this coding level-3. Above level-3 the code is interpreted to produce complete plants and animals including their ability to replicate the whole process in future generations. Coding at level-3 and above is more than a shopping-list of cooking ingredients. It's also the recipe to assemble them into a pie. This is no ordinary pie. It can go on to reproduce itself indefinitely. More than this, coding controls complex behaviour, such as the activity of bees building a hive. The elements of emotional temperament which pass from us to our children lie within the scope of level-3 and above.

The part DNA plays in the manufacture of proteins, is at 'level-2'. This mechanism is reasonably well understood. But the coding of the information needed to make a 3-dimensional creature (level-3 and above) is far from understood. At these higher levels a bottom-up description in terms of chemical bonds becomes too complex to model.

Before any of this mechanism was known, biologists could only take a top-down view of its outcome, the living organisms. If we walk in remote areas we will occasionally come across the skeletons of animals. We can easily spot the same basic skeleton design within a diversity of creatures. Similarities in the rib cages and upper limbs of birds and mammals are striking. The common use of an articulated spine to protect nerves but allow the body to bend and a brain protected within a bony skull. We can hardly avoid using the word 'design'. Exactly the same impression would arise if parts of the chassis of an old truck and an old car were lying side by side. They would both have transmissions, engines, alternators and so on.

climbing up complexity

No two parts would be identical but the overall design would be similar enough to make totally independent development virtually unbelievable.

This kind of visual observation was the basis for Darwin's Theory of Evolution. Similar looking plants and animals were grouped into classes. Different types of 'similarity' were defined based on finer and finer detailed comparisons of individual parts. In the case of extinct plants or animals, fossils were used.

The theory of evolution was first developed without any certain knowledge of its underlying mechanism. Theories of this type tend to develop certain characteristics. Ideas tend to attach to personalities, rivals collect followers. Arguments often become highly personalised. The theory had top-down views on how living things related to each other but was poor on predicting later events from earlier events. Today bottom-up mechanisms are being revealed at an increasing rate. Biochemistry, cell biology and genetics are scientific disciplines at the forefront. These disciplines converged in the mid 20th century. They tend to bypass earlier controversies. They have revealed finer and finer detail of mechanism. It is now possible to match specific amino acid sequences in proteins and compare the DNA sequences in chromosomes.

Overall the evidence supporting the inter-relatedness of all living forms is overwhelming. All of this work has established a web connecting all existing species to each other and to known extinct species. This has been used to show that living creatures share earlier common ancestors.

But even with the superb knowledge of mechanism gained by biochemistry, cell biology and genetics it's just not possible to model the chemistry upwards to predict the complexity of even the simplest bacteria. The test of a scientific theory is its ability to predict outcomes. Using this criteria there is still a long way to go. The top-down theory of evolution is a kind of 'in-principle' account. It's like knowing how the weather works in general terms but still not being able to make a reliable weekly forecast.

The history of botany and zoology is similar to the history of astronomy. The early phases of both were observation followed by the grouping of things observed into schemes of classification. The underlying physics and chemistry were discovered much later. Many instrument techniques had to be invented and developed before the molecular level of insight was possible in biology. These were the equivalents of the optical, radio and X-ray

climbing up complexity

telescopes of astronomy. For biology, optical microscopes and later electron microscopes allowed smaller and smaller organisms to be seen down to an individual virus. The advent of X-ray crystallography allowed three-dimensional molecular structures to be seen. Computers allowed three-dimensional molecules to be modeled more accurately. Again we see the search downwards looking for underlying mechanism: the unbreakable tie back to atomic structure. How small are our physical roots?

(Figure 4.4) complements *(Figure 2.1)* and looks at the range of sizes downwards to the atom.

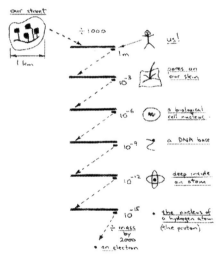

Figure 4.4
Title: 'shrink and repeat'

Visualise tiny distances as just repeated shrinks by 1000 times. We can see 1000 times in one view of a tape measure, 1mm and 1 metre.

Source: own

We come full circle, back to mechanisms at the atomic and molecular level. Once again we come back to the realm of the electromagnetic force. All the tissues of life and DNA itself are constructed by different variants of this force.

Where, when and how did all of this start? How did the mechanisms of self-replication move from the 'elementary' type of replication of crystals (we called level-1) through more elaborate (level-2) enzyme-like chemical templates to level-3 and above, complete instruction sets for living organisms. These are difficult questions. There have been many attempts to suggest possible answers and undoubtedly many more to come. Some amino acids can form in a suitable chemical environment without biological assistance but nucleic acids such as DNA are felt to be too improbable for chance fabrication. Production of some of the basic smaller building-block carbon compounds has been demonstrated in the laboratory.

But the really significant step to large molecules such as DNA, which make their coding debut at level-2, remains unresolved.

climbing up complexity

One idea originating from the 1960's suggested that these elaborate carbon molecules had silicate (clay) precursors. The work suggested that the chemistry of silicates, in a suitable geological environment, could develop early coding processes. The extreme improbability of carbon structures randomly generating coding processes was side stepped by this idea. Later, carbon molecules were able to substitute for parts of the silicate process. The richer chemistry of carbon molecules then led to 'genetic take-over'.

Another idea again originating from the 1960's and still supported, involves DNA's very close relative, RNA. RNA is the genetic material of viruses. It was suggested that RNA had a dual role in early organisms, gene and enzyme. This is made possible by the different 3-dimensional behaviour of RNA with respect to DNA. DNA forms into the famous double helix whereas RNA, depending on its sequence of bases can form many shapes in space. This is the key to an enzyme, the ability to form a 3-dimensional template of suitable shape to bring other potentially reacting molecules and atoms together.

The major problem in all explanations is the step from the carbon compounds which synthesise relatively easily to very much more difficult and elaborate molecular structures such as DNA. Concentration levels need to be high enough to make these reactions more likely and chemical contamination levels which could produce alternative reactions must be kept low. The early chemical formation of isolating membranes, the fatty acid cell wall of the earliest single celled organisms is suggested as one solution to this problem.

In moving from the level-2 to the level-3 stage, the model moves further from a blow by blow, bottom-up account. Even the development of the simplest single celled organism involves much more 'after the event description' than convincing predictive modelling. It is in interesting contrast to current developments in astronomy where more and more computing power has allowed 'bottom-up modelling' to produce better and better matches of large scale events such as collisions between galaxies. The difference in the complexity of these two areas is so profound. Galaxies are at our level-1. Biology is at our level-3 and above, where current bottom-up modelling simply cannot handle the complexity of the system.

After the arrival of single-celled micro-organisms another leap in organisation occurred. Instead of single-

climbing up complexity

cells dividing to make replicas of themselves there emerged cooperative multiple-cell groupings. Specialisation then began inside such multiple-cell groups. Individual cells took on specific shapes and chemistries. The 'standard model' states that all of this development was caused by the preferential selecting out of the effects of chance mutations in DNA codes. Eventually there emerged tiny creatures with mouth-parts, guts, sensory nerves and sexual reproduction to share more mutations. Additional to their physical form they have behaviour patterns, some of which also pass on in coded form.

Further down the evolutionary road the same two mechanisms of mutation and natural selection resulted in all the reptiles, mammals, giant redwoods and so on.

All of this seems unbelievable at first sight and probably at second and third sight. It is argued that this incredible outcome is the sum of an enormous number of tiny steps, each of which is credible. Care is required. This kind of reasoning is very sensitive to abuse. A large number of small gaps can conveniently be ignored or considered closed *'in principle'*. This process can give the erroneous impression of completeness in the top-level summation of all the steps, risking an altogether too self-satisfied model.

The stability of gross features, such as the skeletal structures of birds or mammals mentioned earlier has also to be accounted for. This is generally done in terms of risk. The risk of getting to the totally new compared with the relative safety of reuse and modification. It's perhaps a bit like mountain walking. (*Figure 4.5*)

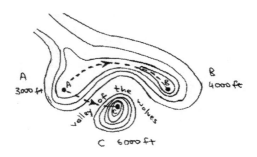

Figure 4.5
Title: 'a performance surface with inaccessible peaks'

From A, journey to B is possible. Journey to C is not. It has been made too costly!

Source: own

Once we have climbed one hill we look for a ridge to take us to the next summit. We want to avoid having to go down to another valley before we climb to the next summit. In evolutionary terms those random mutations which would involve a dismantling of the

climbing up complexity

current structure to allow the build up of say, some new skeletal form, would carry a sufficient immediate penalty for survival, that the new form would never be reached. However, this is another *top-down* 'explanation'. It is a plausible description of mechanism but is incapable of predicting the detailed outcomes we see all around us.

The biggest challenge to the evolutionary model comes from its weakness in giving a bottom-up account of the evolution of the higher level structures in biology. That is the single cell and upwards to the most complex integrated systems such as sight. (In the next chapter, we will glimpse just how complex). This area has attracted new interest bringing in models from other fields of research such as adaptive and self-organising systems.

After pondering on the 'standard model' of evolution we may feel that something must still be missing. From the chemistry at the start of our story, there seems to be an overall theme, complexity increasing and succeeding. This is the reverse of what would normally be expected in physical systems. Of all the possible random mutational changes occurring, most would be backward towards non-survival. But to move forward from a common origin, a viable path forward path must always exist.

The Conscious Universe

Otherwise it would be like finding an exit in a maze that wasn't designed to be a maze in the first place.

Is there a character difference between scientific models for the pre and post-biological phases of the universe? Ignoring the problem of the absolute origin of the universe, we seem to be reasonably confident about the way in which physics can describe the arrival of the pre-biological Earth. We know that lots of pieces are missing or incomplete but we feel that more of the same will improve the position.

The same bottom-up approach in the biological world seems less reasonable. It's only made more reasonable by looking at each small step separately, having accepted the previous step as a staring point, even although it may also be incomplete. But the result of the sum of the steps still seems to be incomplete. How do we build a richer model good enough to predict the complexity of living organisms? A model which will account for life emerging out of forces operating at the molecular level?

The 'standard' model of evolution allows of no top-down 'design' or 'intent', only the relentless process of selection from the effects of random copying errors. Out of this process, have come human beings who are full

climbing up complexity

of 'intent' to do things and full of 'design' to make and achieve things.

What do we make of a blind system which without top-level intent produces at its top level in complexity, a top down planner full of intent? Nature seems to have run her course towards top-down design. By producing us, Nature now has top-down design. The 'standard' theory of evolution somehow seems blind to this cosmic irony. It can provide a general explanation for the evolution of *interactive robots*. However, the extra property of conscious intent leaves a momentous gap in the explanation. It's no answer to simply argue for some kind of competitive performance advantage, in order to explain the evolution of consciousness. To carry any weight, such an argument would need a parallel understanding of how consciousness can emerge out of the evolution of brain matter. We will return to this issue in a later chapter.

Let's return to a biological example for another sight of top-down planning arising out of bottom-up evolution. Through evolutionary time some single celled micro-organisms specialised into multi-celled structures. Later these multi-celled forms emerged with greater cell differentiation forming more complex creatures. Complexity grew with time. Eventually elaborate sensory organs and central nervous systems appeared.

Within the womb a similar complete evolutionary development occurs for each individual member of a species. First successive cell division grows a single cell into an undifferentiated cluster. Then differentiation of cell types, to allow the emergence of skeleton, limbs, and internal organs.

Bottom-up evolution took around 3.6 billion years to produce a human being from a single celled ancestor using random mutation and natural selection. A mother achieves a similar outcome, single cell to human being, in 9 months. That's a time compression of around 3 billion to one. And the difference is a good top-down plan, coded into the DNA of the fertilised single-celled egg.

Evolution is telescoped into each new generation. Echoes of the earlier living forms are still retained as stages in the development of the embryo and foetus. However the stages are now switched on by a plan, rather than waiting for endless mutation and selection. The

DNA carries a record of the previous pathway, found by evolution.

(Figure 4.6) allows a comparison to be made of embryos from fish to human. Significantly, the embryos of different vertebrates are more alike, the younger they are.

The Conscious Universe

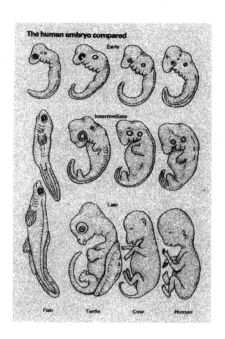

Figure 4.6
Title: **'vertebrate embryos compared'**

The structural similarities across the embryos of fish, turtle, cow and human are greater the earlier the comparisons are made.

Compare a diagonal sequence from top left to bottom right with the right hand column from top to bottom. The similarities eloquently suggest that development in the human womb echoes our evolutionary reptilian origins. Bony fish have been around for 400 million years.

Source: scanned from, *Health and the Human Body, page 21, ed. Dr. Bernard Dixon, pub. Perseus Press, 1986.*
ISBN 0267 396227

3

climbing up complexity

The history of biological evolution on Earth has been built up from different technical viewpoints. This history suggests that organisms of bacterial complexity existed at least 3.6 billion years ago. Life originated no later than about 1 billion years after the planet's formation 4.5 billion years ago.

Over this period life has transformed the atmosphere and surface of the Earth into an enormously interconnected system. At the time of the earliest bacteria-like organisms, the Earth had almost no atmospheric oxygen. When photosynthesis evolved, it introduced oxygen into the environment. This was initially bound into iron-rich rock until significant amounts entered the atmosphere about 2.3 billion years ago. Modern levels of oxygen were not reached until 500 million years ago. Evolution has produced a living planet of plants and animals interconnected through a shared environment of their own making. A tree changes its neighbourhood, reducing light and water, making an environment which will fit some other biological form. A dead tree makes the environment for fungi. Fungi transform the material of a dead tree into nutrients for a living tree. Living animals and plants make the environment for bacteria and viruses. So the simplest life forms live inside the more complex ones. Plants produce oxygen and consume water and carbon dioxide to make up most of their mass. Every niche is explored and filled. Living organisms destructively consume other living organisms but sometimes they interact constructively such as with some insects and plants. This mutual interaction between living organisms, directly and through a common environment creates an enormous level of complexity.

Many loops exist in the system. They contain complicating time delays which produce oscillating behaviour. Over geological timescales large variations, on the scale of ice-ages have occurred. Random events add to an already complex system. It is suggested that the demise of the dinosaurs around 65 million years ago was due to a major collision with interplanetary debris, which put dust into the atmosphere. The current biological face of the planet has been made through such traumatic events, with several mass extinctions. The event 65 million years ago was very significant in determining the current balance between mammals and non-mammals on

climbing up complexity

planet Earth today. These huge disturbing impulses shift the equilibrium of the overall system, which will favour some reactions and oppose others. Taken with the individual self-adaptation of living organisms the whole life-system of the planet is complex in the extreme. Because of this, it's long term dynamics are impossible to accurately predict.

About 14 billion years have passed since the earliest phases of our universe. Around 10 billion years to reach the conditions which could support DNA coding and start the processes of biological evolution on Earth. Because the universe is expanding the passage of time means growth in size. No wonder we live in a universe which dwarfs our Planet. Last century it was popular to use this vast size to suggest the insignificance of humankind. But now this measure has to be balanced against another, that of complexity. On this measure we seem to be at the other end of the scale altogether. We can bring together life, complexity and the basic forces of physics into one revealing graph.

The Conscious Universe

(Figure 4.7) plots the relative complexity of the objects in our universe against their size. Complexity is impossible to define in absolute terms but I don't think many would disagree with the relative ordering of the objects.

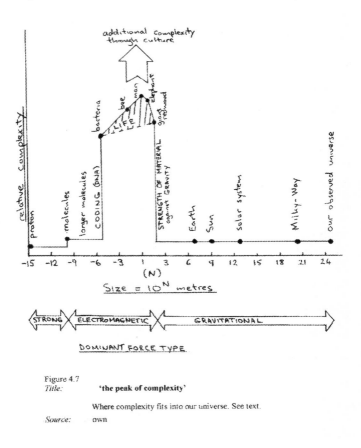

Figure 4.7
Title: **'the peak of complexity'**

Where complexity fits into our universe. See text.

Source: own

The divisions on the horizontal axis indicate size. They are in the form '10 raised to the power 3', so each

climbing up complexity

division represents 1000 times bigger than the division immediately to its left. One metre, about half the size of a human is near the middle.

On the left-hand axis, complexity increases from the bottom to the top. This starts with the basic atomic nucleus, where the short-range strong force dominates. Complexity then progresses upward with the addition of electron shells to nuclei to form atomic elements. Complexity increases again with the atomic elements combining into molecules. We are now in the realm of the electromagnetic force. Increasing size still further, we reach the critical take-off point for complexity. This occurs when large molecules can form a chain which supports coding, such as DNA. The next leap in complexity is huge, with coding enabling molecules to combine to form much larger entities such as bacteria. Moving further up the axis, complexity continues to increase. Structures develop which have the complexity of a human brain and body. Then as size continues to increase, complexity declines a little to the elephant, arguably!

A further increase in size and something very dramatic occurs. There is now a decrease in complexity with increase in size. We reach the giant redwood tree.

As size continues to increase, single living creatures can no longer exist. We might suggest that's because of the strength limitations of biological materials. This is complicated by lots of other factors. First, there is the early 'freezing in' of 'design' features. This might be expected to have occurred in an earlier phase, when creatures were smaller. These 'design' limitations are maybe the real problem. Because of the 'freezing in' effect the 'design' can only modify in a limited way to take care of increasing weight. When a creature gets twice as big its weight increases by eight times (because its volume increases with the cube of its size). However its skeleton only gets four times stronger (because the cross-sectional area of its bones only increases with the square of their size). This is why the elephant has proportionately bigger diameter legs for its size than the mouse. The early 'freezing' of basic design made the six-legged elephant a non-starter! The basic 'design' of the heart and blood system may put a similar limitation on size. Again the volume of blood required to supply tissue increases with the cube of size whereas the capacity of blood vessels only increases with the square of size.

climbing up complexity

Carrying along the size axis, complexity falls again to relatively low levels. Objects the size of the Sun and upward are back to the 'unadorned' basic physical forces which hold material together. There is no coding in the Sun! The gravitational force is in command forcing temperatures up out of the range of complex chemistry.

The Earth is in a slightly different state. Because of life, it has more complex behaviour at a large scale. It has an extended 'eco-sphere' which gives it its unique appearance within the Solar-system. *(Figure 3.4)*

So what can we say about the cultural arrow at the top of the sketch? Could coding again be the take-off point for growing complexity. Spoken and written language, including mathematics, is another manifestation of coding. It took hundreds of thousands of years for the human brain to develop spoken language and perhaps less than five thousand years to develop written language. Language extended thinking and communication. It allowed new information to be spread rapidly without requiring direct experience.

This led to elaborate cultural systems of co-operating individuals.

Interestingly by using this kind of network, nature has found a way to grow complexity beyond the physical limits of individual brain size suggested in *(Figure4.7)*. Language led to civilisation and an explosion in cultural complexity. Civilisation led to specialisation and for a select few more time to theorise about the world. Perhaps less than two thousand years ago something akin to modern scientific methods of discovery began to emerge. They provided the intellectual tools needed to uncover the mechanism of the physical world. Once found, this knowledge was exploited for practical ends. New materials became possible. Machines were designed to make new goods. New techniques were applied to agriculture, animal husbandry and medicine. New instruments extended the range of the senses and so the range of ideas; telescopes and spectroscopes for use in astronomy, accelerators for use in physics, microscopes and scanners for biology. Computers allowed the expansion of scientific models in range, complexity and speed.

climbing up complexity

What next? Let's take a longer view of the evolutionary process. This is mere conjecture but many may still find it disturbing.

Through the action of elementary physical forces acting over a span of 14 billion years, nature has produced complex adaptive organisms. Some have become so complex that they have intelligence and consciousness. Humans use their intelligence so successfully to analyse nature, that some of nature's methods can now be overwritten. Genetic engineering may allow the limitation and possible cure of many inheritable diseases. There are future possibilities beyond disease control, perhaps influencing emotional response, physical or artistic ability, or intelligence. We discuss these topics, not to pass an ethical judgement about them, but to try to imagine a step taken outside of our present time and view.

Consider the precedents. Humankind already reaches out to modify the physical world: making new materials, extracting energy, making machines.

(Figure 4.8) shows some examples of how humankind has already manipulated the mechanics of atoms, molecules and genetic coding. More knowledge has

The Conscious Universe

always been followed by more manipulation. How will this theme flow up the ascending levels outlined in the figure? From this perspective where goes humankind in 1000 years? Where in 10,000 years?

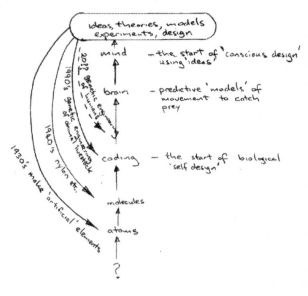

Figure 4.8
Title: **'natures next step in design?'**

Mind adds to and modifies the chain which produced it.

Source: own

Are we now reaching the possibility of another turn of the spiral, designed-in mutation adding to natural mutation? Is the evolution of life on Earth heading to an even larger loop where the physical operation of evolution

climbing up complexity

is taken over by its own outcome, conscious intelligence? Is what we are glimpsing, in the earliest forms of genetic engineering, a hint of a much larger loop back to the genes?

In the next chapter we will look at the brain, the organ which extends complexity beyond biology. It is the agent of cultural complexity. It is the structure with the strangest of property of all.

CHAPTER 5

summit brain

The human brain has about 1 million, million nerve cells. That's ten times the number of stars in our galaxy. The galaxy's large scale behaviour is governed by the gravitational interactions between the stars and by some other not fully understood ingredient, the 'dark matter'. The brain's large scale behaviour is governed by interactions between the neurons. These interactions involve forces which are just the same physical forces as drive the chemistry of everything else we can see in the universe.

(Figure 5.0) shows a microscopic view of a neuron within the cerebral cortex of a human brain. The finer fibres, the dendrites, carry electrical input signals to the cell body and the thick fibre, the axon, carries the cell's electrical output. The output of one cell connects to one of the inputs of another, making

up a vast network. The role of the neuron is to send an output string of electrical pulses down the axon, when a particular combination of inputs triggers the cell. The output pulses are the result of rapid changes in the concentration of ions across the cell wall. The neuron uses at least four different types of ion; sodium, potassium, calcium and chloride. Once initiated, the pulses of electrical charge propagate along the length of the axon. The concentration of ions is controlled by pores or ion channels made of specialist proteins (again that most versatile of materials). The cell then slowly pumps these ions back to the outside of the cell. The neuron cannot transmit again until all the ions which leaked in, are pumped out. Around 99% of neurons are in the brain. The remaining 1% is distributed throughout the body in the peripheral nervous system. There are many types of specialist neuron. Some output pulses when mechanically stimulated, such as those for touch in the skin or for hearing in the inner ear. In smell, the presence of particular molecules at the sensory neuron cell wall will trigger the pulse stream into the brain. In sight, light falling on the surface of a retina cell within the eye will trigger the pulses down the axon.

This is the lowest level of the brain's function. What is here, to turn this perfectly universal chemistry, into conscious feeling? If we feel we can't answer this question, then perhaps we had best look further up the chain. Let's start at the top, with the whole brain.

The human brain weighs around 1.4 kg, about 2% of adult body weight. However it uses about 20% of the bodies oxygen supply. It is a favoured organ, best protected and given the biggest food allowance. As an organ the brain has developed over hundreds of millions of years. Fossils don't yield much evidence concerning the evolution of brains. Unlike skeletal remains, soft tissues don't generally fossilise. Brain volumes of extinct animals can be estimated from the size of skull fossils but it is virtually impossible to gain any detailed knowledge of their function. It is therefore necessary to look at the brains of living species and make inferences about their ancestors. This is an unreliable process for animal lines which show

summit brain

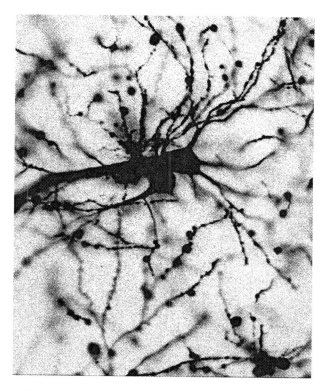

Figure 5.0
Title: **'inside the cerebral cortex'**

Neurons of the pyramidal cell type within layer 5 of the 6 layer cerebral cortex.

Source: scanned from, The pyramidal cells of the cerebral cortex
Atlas of the Body and Mind, bottom page 68-69, ed. *Claire Rayner*, pub. *Crescent Books, New York, 1976*
ISBN 0517 051451

high rates of change in the fossil record but not so unreliable for those exhibiting little physical change through time. Present belief is that the brains of current lizards are not too different from those of their extinct ancestors. Similarly current small mammals, such as bats, probably continue the general brain structure of the fossil bats. The standard evolutionary model suggests that the reptiles preceded the small mammals and the small mammals preceded the larger mammals such as us. Putting all this together allows us to see a record of the evolution of the brain by looking at the brains of currently living reptiles and mammals.

Such investigations lead to the view that the earliest parts of the brain are in the brain stem. This is the largest region (as a fraction of the whole brain) in the brains of reptiles. Relative to the whole brain small early mammals have a relatively larger limbic system. More recently evolved modern mammals have a relatively larger cerebrum and cerebral cortex, again measured as a fraction of the whole brain. The major regions of the human brain are illustrated in *(Figure 5.1)*.

summit brain

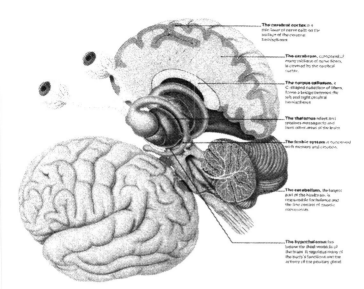

Figure 5.1
Title: **'the regions of the brain'**

Travelling inwards from the skull: cerebral cortex, cerebrum, corpus callosum, limbic system, thalamus, hypothalamus and brain stem; then rearwards to the cerebellum.

Source: scanned from 'The regions of the brain', *Atlas of the Body and Mind*, bottom page 75, ed. Claire Rayner, pub. Crescent Books, New York, 1976
ISBN 0517 051451

The Conscious Universe

In our own brains we have an additional 'view' which is unique. This is the 'inside' view, available to every owner of a brain, comprising sensations, emotions and thoughts. In open-skull surgery on conscious patients, stimulation of different parts of the brain causes the patient to experience different sensations. Since they are conscious, they can be described by the patient. These kinds of observation were originally made around sixty years ago *(Figure 5.2).*

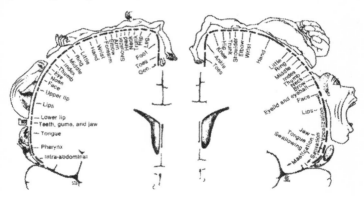

Figure 5.2
Title: **'mapping the surface of the human brain'**

The surface of the cerebral cortex has two adjacent strips, running from top to bottom, on each side of the brain. One of the strips deals with incoming sensory inputs and the other with outgoing motor commands. The diagram shows one strip of each type. A sensory strip is shown on the left and a motor strip on the right. The area of cortex devoted to different parts of the body is drawn around the outer surface of a vertical slice through the brain, from ear to ear.

Source: scanned from 'Somatopic mapping in human....'. The Human Brain, 4th edition fig 3-28, page 72. John Nolte, PhD, pub. Mosby Inc., 1999
ISBN 0 8151 8911 7

summit brain

This shows the mapping of sensory and motor nerves from the body to the brain surface. More recent surgical investigations and treatments give access to parts of the brain less accessible than the surface. Electrical stimulation of parts of the limbic system produces emotions in the same way that stimulation of parts of the cortex produces the sensation of touch. Other direct physical information, linking brain activity to brain region, has been gained from investigation of patients with localised accident damage, or damage caused by diseases such as a brain tumor. Much new information comes from various types of brain scanner *(Figure 5.3)*.

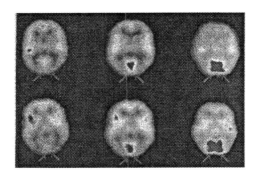

Figure 5.3
Title: **'scans through the brain'**

The two images on the left are horizontal sections through a brain at rest with closed eyes. The centre images are taken with the eyes open and the images on the right are when watching a complex scene. In these scans, short term radioactive 'labels' are chemically attached to the blood glucose and detected by the scanner.

Source: scanned from page 67, *Health and the Human Body, page 21*, ed. Dr. Bernard Dixon, pub. *Perseus Press, 1986*.
ISBN 0267 396227

This information shows that many of the brain's abilities are not located in one place but are shared across different regions. Two-way interaction occurs between these regions rather than unidirectional flow, each region contributing some extra property to raise the level of performance of the overall function. If we summarise the functions of the different regions of the human brain, starting with the oldest in evolutionary terms, then we get an interesting insight into the evolution of our mental abilities. This also gives us a basis from which to speculate about likely levels of consciousness in other animals.

The brain stem
(Figure 5.1)

The brain stem links the cerebellum, thalamus, hypothalamus and limbic system with the top of the spinal cord. Motor nerves, which carry impulses to drive muscles, as well as sensory nerves for touch and pain route through the brain stem. However the brain stem is much more than a conduit. It is home to the neural structures which control blood pressure, heart rate and breathing, the basic body control processes.

The brain stem is involved in determining levels of alertness to the external world, ranging from intense concentration through to unconsciousness and sleep. Some cells in the centre of the brain stem connect to sensory inputs in different areas of the body. Something to remember, next time you're shaken out of sleep.

The thalamus
(Figure 5.1)

Sensory information relating to sight, hearing and touch is first routed to the thalamus, from where it is passed on to areas of the cerebral cortex. Smell is the only sensory pathway not routed through the thalamus. Motor activity also passes through the thalamus with connections to the cerebral cortex and the cerebellum. 'Feedback' pathways exist through which information from a later part of a processing chain is connected back to modify the processing activity of earlier parts. In the case of visual pathways, less than 20% of the neural connections to the thalamus are sensory optic fibres from the eye whereas around 50% are connections returning from the visual area of the cerebral cortex.

The cerebellum
(Figure 5.1)

The cerebellum is one of the larger regions and weighs about 140 grams. It is dedicated to the coordination of movement and balance, covering anything from quietly sitting still, to playing squash, soccer or golf. You may feel it is therefore a very important region. The cerebellum has a folded outer surface to increase its effective surface area, a technique also used to extend the area of the cerebral cortex. Layers of neurons lie in the outer surface of the cerebellum, interconnecting through a network of fibres lying below its surface. Demands for physical movements which emerge in the motor area of the cerebral cortex, are refined by the cerebellum. It stores learned patterns of movement. It enables fine precision of movement by taking inputs from sensory neurons in muscles which give position information to enable a fine-tuning form of feedback. The cerebellum also helps us balance by taking position information from the canals of the inner ear.

The hypothalamus

(Figure 5.1 and Figure 5.4)

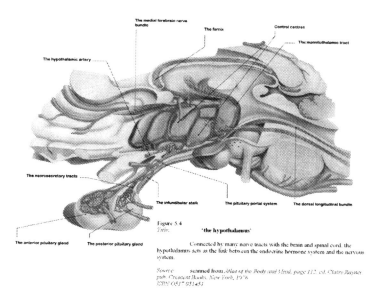

Figure 5.4
Title: 'the hypothalamus'

Connected by many nerve tracts with the brain and spinal cord, the hypothalamus acts as the link between the endocrine hormone system and the nervous system.

Source: scanned from *Atlas of the Body and Mind*, page 112, ed. Claire Rayner, pub. Crescent Books, New York, 1976
ISBN 0517 051451

The hypothalamus is focussed on functions which defend the survival of the organism. Its 'internal' function is to stabilise the environment of the body, factors such as body temperature. Its 'external' work is involved in modulating the feelings which drive survival actions such as hunger, thirst and the 'fight or flight' response to danger. It also influences sexual drive, which is about the survival of the species. The hypothalamus connects the neuronal system of the brain to the hormonal system of the whole body. The gateway of the connection is the pituitary gland which sits on the base of the hypothalamus.

The Conscious Universe

This gland secretes hormones into the bloodstream to control functions such as the amount of water retained in the body. Hormones injected into the bloodstream will flow into the brain and bathe the neuronal system, so completing another potential feedback system. The complete hypothalamus only weighs four grams.

The limbic system

(Figure 5.1 and Figure 5.5)

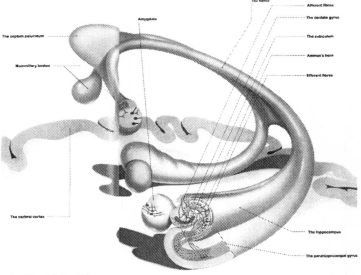

Figure 5.5 'the limbic system'

The limbic system is centrally involved in emotional responses as well as memory and learning. It communicates with the processes occurring in the cerebral cortex.
Source: scanned from, The centre of emotion and memory (the limbic system) *Atlas of the Body and Mind, bottom page 116-117*, ed. Claire Rayner, pub. Crescent Books, New York, 1976. ISBN 0517 651451

The limbic system's functions relate to learning, short-term memory and to emotion such as aggression or pleasure. Direct electrical stimulation of a human limbic system during surgery with a conscious patient produces a range of reported emotions such as 'rage, agitation, anxiety, elation, excitement, colourful visions, deep thoughtfulness, sexual interest and relaxation'. We saw above that the hypothalamus is also connected with 'fight or flight', sexual drive and emotional state. The limbic ring and the hypothalamus are interconnected through the amygdala.

The cerebrum and the cerebral cortex
(Figure 5.1)

In humans, the biggest part of the brain is the cerebrum. It surrounds the space occupied by the other major components. This has allowed it to expand with the least disturbance to the older brain. The cerebrum is home to the most massively interconnected network of neuron fibres in the brain. Most of the neuron bodies lie near the outside of the cerebrum, in the layer of the cerebral cortex. The cortex has many folds in its construction, enough to multiply its area by about thirty times. Processing of incoming sensory information, its storage and outgoing motor commands all happen in the cerebrum. Human

vocabulary is located in the cerebral cortex. Nouns and verbs
(things and actions) are held there in separate adjacent areas. Second languages learned later in life are held in separate places whereas bilingual infants have common areas for nouns and verbs.

If we build a summary which focuses on our specific interest in feeling and consciousness, then we have....

The cerebrum and cerebral cortex connect with thinking and using language and pictures.

The hypothalamus and limbic system connect with drives and the effect of their satisfaction or denial in terms of felt emotion. They are involved in feeling pleasure or pain. The limbic system can respond to stimulation originating from non-verbal drives such as thirst or sex. It can also be affected by 'thoughts' in the cerebrum – cerebral cortex.

The brain stem connects with the level of arousal.

summit brain

Drug therapy can offer another insight into the relationship between region and function (*Figure 5.6*).

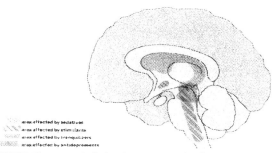

area affected by sedatives
area affected by stimulants
area affected by tranquilizers
area affected by antidepressants

States of consciousness can be altered by the use of drugs. Antidepressants (blue) affect the midbrain, the area involved with the formation of mood. Thus they can prevent and help to cure depression. Stimulants (striped), such as amphetamines, act on the reticular activating system (RAS) and the hypothalamus. Their effect is to decrease fatigue, increase alertness and elevate mood if the mind is overactive and, therefore anxious, tranquilizers (green) may be used to reduce the level of anxiety. They act principally on the limbic system and the RAS. Sedatives (light brown), for example barbiturates, are a group of drugs often administered as sleeping pills. They act on the RAS and on the cortex to reduce the conscious experience that would normally induce wakefulness

Figure 5.6
Title: 'drugs.....direct interference in the mechanisms which underpin personal conscious experience'

Source: scanned from, States of consciousness (illustrates where specific classes of drugs operate)
Atlas of the Body and Mind, bottom page 197, ed. Claire Rayner, pub. Crescent Books, New York, 1976
ISBN 0517 051451

This relates sedatives, stimulants, tranquilisers and antidepressants to their region of action.

If we look at the interconnections between the regions we can glimpse some of the architecture which supports our mental life. We already saw that the limbic system and the hypothalamus interconnect through the amygdala. The limbic system is also in two-way communication with the cerebral cortex. The 'thinking' activities of the cortex influence the 'mood' processes of the limbic system whilst the 'mood' states of the limbic system influence the 'thinking' processes of the cortex. So we have three interconnected brain regions, the hypothalamus, the limbic system and the cerebral cortex. It seems that the hypothalamus is involved in 'driven' behaviour and the limbic system is involved in the subjective feelings resulting from satisfying or not satisfying drives. The limbic system, in its turn, is influenced by and influences the 'thinking' processes of the cerebral cortex. In this way the limbic system in some way allows 'thoughts' to modulate basic drives.

This is all very curious. A processing activity in the cerebral cortex, which corresponds to a thought, affects the state of the limbic system which in some way results in us feeling an emotion. However, when we try to catch ourselves thinking, we often seem to be using words. Sometimes we feel tiny involuntary tongue twitches accompanying a thought. Other times, we speak aloud to help ourselves think. We may think of something which makes us smile and have a feeling of wellbeing. How do words, which are arbitrary and different in different cultural languages, connect with the emotional activity of the limbic system? This system is so much older in evolutionary terms than the modern language-capable cerebrum. The *learned* content of the cerebral cortex has to be mapped onto the emotional activity of the limbic system. Perhaps there has to be some other intermediate level of representation of our thoughts, shared by all humans, in a more 'universal' code than a cultural language. Some form which is already connected to the limbic system. Some theoretical linguists have long suggested the existence of such a genetically based language platform.

The evolutionary development of the brain through reptiles to mammals, including ourselves, was outlined earlier. The part played in human emotion by the limbic system and hypothalamus make it interesting to ask questions about the emotional life of other animals, which also have these structures. Most owners of pets would testify that their charges have a considerable emotional life. In large mammals such as dogs; excitement, contentment, affection, anger and fear, all seem present. The human extra-large cerebral cortex is a late arrival in evolution. It processes and organises information. It is where ideas flow. Perhaps its size is not central to the basis of consciousness. The human cerebral cortex adds the ability to represent ideas in a rich language. By introducing rich ideas, perhaps it only extends the set of objects of which to be conscious. Perhaps our basic emotions can be triggered by a much more sophisticated repertoire of events but are are not anymore intensely felt than by many other mammals, at least. It is interesting to speculate. These questions are central to the ethics of our treatment of other animals. Perhaps knowing the striking similarity in brain structures between us and other animals should raise the importance of these issues.

We will take a final, major system view of the brain. This will show how regions interact to perform a total function. The vision system seems best for our purpose. We can detect the sun through temperature sensors in our skin but only our eyes, unaided, directly connect us to objects more distant. The vision system occupies a larger part of the cerebral cortex than any other sense. It lets us see in colour, in three dimensions and enables us to track moving objects. The system can store to memory and correlate between memory and what we currently see. This is selective enough to allow us to spot the face of a friend in a crowd. The above describes the system looked at from the 'outside'. From the 'inside' the whole visual system provides us with our personal experience of 'seeing'. There are more neurons devoted to sight than any other sense. The human eye has a huge range of light sensitivity and can operate over light intensity levels of one million, million to one. It is this property which enables our unaided eyes to see across 2.4 million light years to the Andromeda galaxy.

Within the eye, images fall on light sensitive cells in the retina. Some of these cells are only sensitive to intensity, others only to reds, others to blues and others to greens. Arranged in layers above these light sensitive cells are other interconnected processing cells, which in their turn connect to ganglion cells. It is the axons of these ganglion cells that are bundled into the optic nerve trunk from each eye.

The left side of *(Figure 5.7)* shows how the fibres within the optic nerves from each eye are split and take different routes into the brain. The route depends from which half of the retina's surface they emerged. The left-hand side of the retina of each eye is paired into a single bundle in the optical chiasm. This bundle is routed to the left side.

The right-hand side of the retina of each eye is routed to the right side. Since the images on the retinas are inverted top to bottom and left to right by the lenses, the left side of the brain receives information about the right-and visual field and visa versa.

These left side and right side nerve bundles now route to the left and right side of the thalamus, on their way to the cerebral cortex. This area of the thalamus is called the lateral geniculate nucleus or LGN. *(Figure 5.7)* shows a horizontal eye-level section through a brain. The bottom-right of the illustration shows a MRI image of a living brain given a flashing visual image. This horizontal scan is through a similar section. The red and yellow areas show activity in the LGN's and in the visual cortex.

The retina has several specialised processing layers. There are outputs which carry information on shape and colour and outputs which carry information about movement and contrast. These specialised functions are retained in equivalent separate layers within the LGNs.

summit brain

Figure 5.7
Title: **'the vision system'**

The left and right eyes' view of the left-hand half of the visual field goes to the right cortex, and vice versa. Within the right visual cortex (and equivalently within the left), the information from each eye is kept in separate layers of the cortex. Specific features such as shape and colour, movement and contrast are also kept in separate layers. These two sets of layers are interleaved like a sandwich to allow short cross connections that underpin stereoscopic vision.

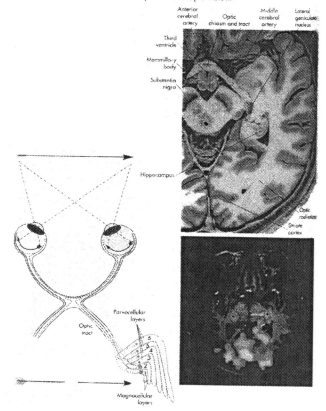

Source: scanned from, figures 17.22, page 417 and figure 17.23, page 418 and parts combined. The Human Brain, 4th edition, John Nolte, PhD, pub. Mosby Inc., 1999
ISBN 0 8151 8911 7

The Conscious Universe

The right side LGN receives nerve fibres from the right side of both retinas and the mirror of this for the left side LGN. These different eye sources within one nerve bundle are also kept separate in the LGNs by mapping each into different layers of the six layer deep LGN. The simultaneous need of keeping both feature specialisation separate and the signals from each eye separate is solved by means of an ingenious sandwich structure. Layers 6, 4 and 1 of the LGN sandwich are for the left eye and layers 5, 3 and 2 are for the right eye. Layers 3, 4, 5 are dominant in shape and colour and 1 and 2 in movement and contrast. This link-up puts the different sources in an intimate geometrical relationship.

Nerve connections radiate out from the LGNs to connect to the visual part of the cerebral cortex at the rear of the brain. 'Next neighbour' geometrical mappings are preserved in the cortex as well as separate layers for specialised processing. The visual cortex is organised in columns, perpendicular to the cortex surface. Adjacent columns are associated with adjacent points on the retina, which correspond to adjacent points in the visual image. Each column has specialist processing taking place in different layers, relating to shape, colour and motion. If a particular column takes most of its information from the right eye, then its adjacent neighbouring column will take its information mainly from the left eye. The adjacent columns communicate sideways to bring together all the left-right information about a small region of the visual field projected on to each retina.

This information is then passed on to higher visual processing centres in the cerebral cortex. More networks of interconnected neurons. The same general structural basis is repeated everywhere. No extra ingredient, only different networks.

We now have the top-down description of the visual system given above to add to the bottom-up description of the basic electrochemical components, the neurons, given earlier. However, are we any further forward in understanding the step which turns all of this into our subjective awareness of seeing? The enigma remains. How does all of this mechanism generate our *conscious sensation* of being in a colourful, seamless, 3-dimensional world. Where is this *experience* engineered?

The more we investigate, the more questions are uncovered which demand more investigation. The study of the brain illustrates what is found everywhere in the biological world. There are layers and layers of mechanism with each layer showing wonderful elegance and intricacy. The deeper we delve into underlying mechanisms, the more we return to chemistry and physics. How the molecules are held together, why they are shaped the way they are, why a particular ion passes through a particular ion channel. If it is not possible to model these aspects completely, then we may incline to feel that the difficulty is complexity, rather than principle.

The 'down' direction uncovering these layers of underlying mechanisms has yielded results to tenacity, better tools and time.

The 'up' direction, accounting in detail for the emergence of layer upon layer of organisational complexity, has also yielded to tenacity, better methods and time. But it still feels very much the weaker direction.

It's like painting. An artist thinks up a scene and takes the paints from his box and paints an image of this scene.

A visitor arrives. Visitor 'number one' looks at the scene in the painting. He is interested in what material has been used. He scratches around on the canvas and soon has samples of the different pigments. So visitor 'number one' knows the picture produced and has now found out something about the material lying below.

A second visitor arrives after the painting has been put away. 'Visitor two' only sees the artist's paint box. He quickly knows all about the material used in the production of the picture. He will never be able to reconstruct the picture.

Does this echo our central question? Can the visitor 'science' ever explain the existence of consciousness from the kind of material she knows?

Now that we have tried to know something of the material we are ready to focus on the problem of explaining consciousness.

CHAPTER 6

the hidden outer world

Let's bring together our excursions into astronomy, biology and brain.

We are tiny (one thousand million million times smaller than our galaxy) and we are huge (one thousand million million times bigger than a hydrogen nucleus). We are made of universal material interconnected with massive complexity. Each of us is located inside a brain. Our brain is connected to sensors which relate to sight, sound, smell, touch and taste. The brain/mind perceives these and translates them into experience. We are individual centres of feeling. Thus the lifeless things surrounding us are given personal meaning. Radiating energy is enjoyed as sunshine. Vibrating molecules are as pleasing as hot tea.

the hidden outer world

After feeling and consciousness have evolved, emotion is possible. With emotion, fear, curiosity, mystery, kindness, love, joy, disgust, sadness and awe become part of our experience: everything that makes us human.

A huge amount has been written and spoken about the items contained in this list. But it is not necessary to enter such wide-ranging territory to suggest why there might be a fundamental difficulty in the way of a scientific 'theory of everything'. We can investigate more basic sensations such as the feeling of pain or the sensation of seeing red or feeling hot.

We are in contact with the world through our senses. We feel these senses as vividly separate experiences. These divisions have been made by evolution to nable us to eat, survive and reproduce. They probably correspond to the 'outer world' only in as much as they group 'outside' information in the most efficient way, to get the best advantage out of the development and maintenance cost of a brain.

At the lowest level our sensations have fixed how the 'outer world' must look to us.

Let me try to clarify with two examples. The first involves sound and touch. The second involves sight.

What is the difference between sound and touch? We feel the world of sound to be different from the world of touch. The wind blows on our skin and on our ears. Similar mechanical movement of molecules makes us feel sensations of both sound and touch. If the nerve ending is in our inner ear it is sound. If the nerve ending is on the surface of our body it is touch. It is the region of the brain connected to the nerve ending which determines the type of sensation we experience.

Where is whiteness? We have probably all seen the demonstration of the primary colours. Three coloured lamps shine on a screen, red, green and blue. Where they all shine and overlap looks white. What is happening? The red, green and blue light are all reflected off the screen into our eyes and projected by the lenses of our eyes onto our retinas. There the three colours stimulate three types of cone cell. One type is sensitive to red, another to green and another to blue. (There are also rod cells which work in the blue/green region of the spectrum. They don't play a part in light brighter than moonlight.) We saw in chapter 5 that electrical signals from these cone types are separately routed through the thalamus to the visual cortex of the brain. Only there, are they integrated. The consequence of this is that we see white. So white is not a property of the world 'out there'. White is a sensation constructed within us. Evolution has produced this outcome. Perhaps we could have had a non-integrated colour vision. The three colours could have been kept separate in some totally different form of visual perception. Three separate colour spaces running in parallel just like hearing and touch and smell. The evolution of the visual system found integration a better path. We can see different colours as long as they are spatially separate. But if different colours come from the same direction they must be added to produce a single perception. So colours are integrated at a 'lower level' than vision is with hearing for example and cannot be unbundled.

Evolution's balance of benefits over time now determines how we perceive parts of the world to be white instead of composed of separate colours.

All of our conscious sensations lie behind this type of screen or filter.

Do we have similar predetermination in our immediate sense of time? Think of 'now'. In our immediate experience of 'now' the world seems to flow past us. Things which happened a fraction of a second ago seem in some sense still present with things happening closer to 'now'. But things that happened a few minutes ago, we have to recall in another way. When we make an effort to do that, we seem to be able to run them again with a similar feeling of short-term flow.

Think about the way in which the length of this 'now' time window affects how the world looks to us. The duration of 'now' must depend on some mechanism of very short-term memory. Let's use a convenient analogy, such as a video-tape. For a fixed amount of memory storage we could record over a much longer time window if we ran the tape more slowly and didn't try to record or replay events which changed rapidly. We have this sort of limit already. It's what stops us seeing the spokes of a moving bicycle. We could push this to lower and lower speeds and remember events over a longer period.

Alternatively, for the same memory storage we could shorten our 'now' time window, speed up our sensors and recording speed, and see the individual flaps of a bee's wings!

the hidden outer world

Or we could increase the amount of appropriate memory in the brain, keep everything else the same and make 'now' longer!

Presumably evolution has traded all the factors to fix the length of our personal 'now'. It may have been optimised for interactions with moving objects when our evolving antecedents either hunted their prey or ran from their predators. The speed of prey or predator, determined by their mass and power, may have determined how the passage of time looks to us now: another built-in filter.

Are there even more built-in screens and filters operating at a level above the immediate senses? Our perceptions about the world contain certain pre-dispositions. In chapter 4 we suggested that it is an inherent human characteristic to put 'purpose' into everything we see. We have ears so that we can hear. However, the 'standard' theory of evolution doesn't support suggestions of 'before the event' purpose or intent. Phrases such as… ' 'x' happened so that 'y' would be possible'… don't fit into the formal theory. It suggests that things worked out by the summation of random mutations in DNA selected out by survival rate. It suggests that the hugely complex vision system (described in chapter 5) has been built by natural selection from lots of small incremental changes without any large-scale direction. As designers and planners, ourselves, it's difficult to comprehend how this level of coherent complexity has arisen without 'design'. There is a mix up between our formal science and our everyday habit of seeing purpose in the forms of the world. Can we see where up-front 'design' could reside inside any scientific account of the world? Design relates to 'intent'. It involves more than a set of forces which will produce a future result. Design implies intent, where the shape of the outcome exists ahead of its arrival.

Is there a sense in which the chemistry of the embryo carries a form of 'intent'? An 'intent' not possible in the chemistry of the early universe. The DNA recipe in the fertilised egg 'uses' chemistry to produce an intended outcome. The intended outcome is drawn in the DNA ahead of the process of making the embryo. This notion of design and intent only seems to fit into systems which have coding. That is systems which have ways of representing outcomes ahead of their arrival, systems which have encoded plans and recipes. There is no equivalent representation of outcome ahead of action in pre-biological chemistry or physics. There forces produce outcomes which are not to found in any encoded form.

the hidden outer world

I think it would be useful to summarise this view of 'design' and 'intent'. Can it help reduce the discomfort some of us feel when faced with the 'standard' theory of evolution?

Consider three discernible phases in the arrival of design in our universe.

1 The universe before biology, that is before coding.

This is a universe of physics alone. It is not engaged in design. This universe doesn't experiment with itself. It is engaged in construction. The existing forces express their actions 'directly' on the content of the world. It is a world of atoms, molecules and stars.

2 The universe at the arrival of biology, that is at the arrival of coding.

This is a universe of physics plus design. How can we use the word 'design'? DNA carries in its structure information for future use. It carries a recipe which when enacted will produce a future product. It's like a material list and a set of instructions to build a house. We wouldn't balk at using the word 'design' in the case of the house because we can imagine an architect. The organisation of the DNA recipe has come from a process of 'self-design'.

The recipe has been written in a form which allows experimental change. Changes are then tested by natural selection for improved or diminished performance.

The architect has been replaced by a copying machine which sometimes makes errors when copying the plans. The builder just builds and the customer decides whether they like the change. The current plans and instructions sit inside the DNA, ahead of the next construction project. Is calling this tested accumulation of changes a 'design' too inappropriate?

3 the universe today

This is a universe of physics plus biology plus feeling and consciousness. It certainly has design. We design at a level 'higher' than occurs in the universe of (2) above. We 'design' in the realm of ideas and models. We can experiment without commitment to action in the larger world. Because of this, design has been freed from its inertia. It can create and investigate new forms millions of times faster than the universe of (2).

As humans we add conscious intent and design as a layer above the non-conscious experimental 'design' activities of DNA. However, on reflection, perhaps we

the hidden outer world

sometimes resort to a design process very like DNA's. Think of an aircraft designer who makes prototypes with slightly different wing shapes. He tests them in a wind tunnel to decide which ones to keep. Is the keeping of a current plan up-to-date, with successful experimental changes, not exactly what DNA does?

Recall the vision system described in chapter 5. Is it credible that such a system, integrating so many properties such as colour and stereoscopy, could come from the sum of small disconnected individual random mutations? On the way to stereoscopic vision, how would an individual mutation offer enough advantage to be selected? If the individual advantages were not related to stereoscopy, why would all the separate steps add up so neatly as a system? We need the division of the retina, the separate routing of nerve tracts and the sandwich structure of the LGN and visual cortex.

Is there something yet to found which directs the evolution of complex systems? Is there some sort of non-conscious 'design' in the sense we outlined above but operating at a higher level? Is it possible that there are real layers of encoding in the genome, which allow the separate, largely independent, evolution of organs and other features. Can we ever find such a high-level language written in the low-level alphabet of DNA?

Let's try to illustrate the difficulty with a practical analogy. Imagine running an application level program on a personal computer. Take the example of a chess game. The top-level of the machine presents the game to us, in a natural language such as English and probably gives a graphical display which we can easily understand. At the bottom-level of the machine all this is achieved through the movement of electrical charge inside pieces of silicon. In between these 'levels' are several layers of computer languages which are translating up and down a hierarchy of levels. These levels are present to match the lowest level to the highest in a way which allows maximum reuse of the lower levels. That is why a PC can run so many different high level applications. It can be a word processor, a mathematical problem solver or a chess game. Every application reuses the same lower level parts. Does this remind you, just a little, of the reuse of protein molecules, DNA codes and skeletal forms in different living organisms?

the hidden outer world

One day we start playing another game of chess. Suddenly we feel especially inquisitive. How does our PC actually play chess? At the top 'functional level', we know all about chess; we have no difficulty understanding what is going on at this level. If we spent long enough and had the correct 'tools', equivalent to the tools of our earlier astronomer or biologist, we could dig down and discover that below the level of chess there where other processes going on inside the PC. We would discover that there were little black box-like objects with lots of metallic legs, all connected through long strips of metal. Using one of our instruments, we would see electrical pulses streaming along these strips. We would find repeating patterns in the pulses. We might infer they were characters in some sort of coded language. If we broke open one of the little black objects we would find an even smaller world of tiny objects all connected up on an even more dense scale but again never any sign of chess. Further on in our investigation we would discover that the most basic activity for the inside of these little black boxes seemed to be the complicated exchange of electrical charge. Everywhere, patterns of moving electrical charge.

We could play a few games of chess, whilst observing this strange electrical activity. But it would be hard to believe that we would ever get beyond building a model of this behaviour which ran along the lines……..

'The chess abilities of this machine are the outcome of electrical activity within its smallest parts.'

Does this remind you of our knowledge of the genome? To do any better, we would need some prior knowledge of the internal hierarchical levels, between the two extremes of chess and charge. We would need knowledge of the content and language of the high-level programming layer of the machine. We would need knowledge of the languages below, which translate down. This would eventually reach the level which translates into the movement of electrons, at the lowest level. The chance of finding these intermediate levels from observations only of chess and charge is remote. They might never be discovered. More optimistically, we might *invent* one of many possible hierarchies which could explain things to a certain degree.

This example has been used to illustrate one difficulty. That of finding the intermediate layers of a complex process when they actually exist. We know they exist because they were put there by humans acting as

the hidden outer world

designers. The physical hardware and then the software to run on top of the hardware, each level had it's own group of experts.

In the equivalent biological case, *we* are the top functional level. Only the lowest levels are so far understood. They are the levels at which DNA codes for basic materials such as proteins. The current level of decoding which is possible for the human genome, is akin to the level of our inquisitive chess player. He knows a lot about chess, a little about electricity and virtually nothing about software.

As it is, the existence of even the simplest life forms, such as single celled animals, cannot be predicted from chemical models. However, we quickly slip into using *purposeful* language. For example, 'the cell has a membrane to give a protected chemical environment'. We insert the logic needed to produce satisfying explanations. This overlay of everyday purposeful language can obscure just how strange the 'standard' model is. It disguises its weaknesses and makes us feel more comfortable.

All sciences use intermediate conceptual levels. Some turn out to be illusory and are only kept because they simplify our thinking. We saw this earlier, with gas pressure. This is a simplifying idea which removes the problem of thinking about the motion of each individual molecule which makes up the gas. However, what does nature do? Nature *will* be operating on every individual molecule, all of the time.

Common-sense makes us feel that some intermediate boundaries definitely exist, such as the organs of the body. However even here, a little reflection makes things less definite. Consider the organ we call the heart. Where does it begin and end? We might cut through the arteries and veins connecting it to the rest of the body and say, 'well let's make the boundary here'. However there is no natural division at such a cut. The arteries and veins connect onwards to capillaries in the body's tissues. The membranes of the capillaries pass components of the blood fluid into the intercellular body space. From there the fluid reaches the membrane walls of the body's cells. The divisions we introduce and *name* are for our own convenience.

In summary, our perception of the 'outer world' is restricted by the evolutionary history of our senses and basic mental structures. This filter is fundamental and impossible to circumvent.

CHAPTER 7

limits to intuition

In Chapter 6 we described some fundamental barriers between us and the outer world….the world which still exists when no one is there to see it. We cannot see the outer world except through the preconditioned filters of our basic senses and ideas.

At some very basic level we have ways of picturing the world. Ways we have biologically inherited and are stuck with at the root of our higher-level thinking. These built in propensities presumably evolved because they gave a survival advantage within the outer world. However, when we enter new environments they may become inappropriate. The shared root of physics and everyday physical intuition is hinted at in the letters M.K.S. These stand for units of measurement in the international system of science. 'M' is the metre for length, 'K' is the kilogram for mass (let's treat this loosely as quantity of matter) and 'S' is the second for time. Other measures such as speed and acceleration derive from length and time. Now common sense intuition feels independence here, for example we can change speed without changing the others. For centuries science shared this seemingly self-evident belief. As physics developed it extending its range of interest from objects on the scale of our everyday experience to objects which were immensely bigger or infinitesimally smaller or enormously faster. Nothing disturbing would have happened if the universe behaved the same way at all scales. But this was not to be. All was well until the scale of what was being observed moved far outside of the range our intuitions had optimised to cope with.

The survival test for science is its ability to make correct predictions. Failure to pass this test, early in the twentieth century caused physics to depart from its alignment with common sense intuitions. Time, length and mass became interdependent in Einstein's Theories of Relativity. Cause and effect lost their direct intimacy with the coming of Quantum Theory.

limits to intuition

By the middle of the 19th century, the 17th century ideas of Newton had been applied to all the scientific problems of the age. Mechanical phenomena such as the motion of bodies interacting through gravity or by direct collision were well accounted for. Embedded in Newton's ideas were 'action' and 'reaction'. Apply one force and a reaction will occur. A second thing is caused by a first thing. Much of the mechanics of the known universe could be described within Newton's system of rigorously linked causes and effects. The effect of a present cause became the inevitable cause of a next effect, endlessly. Since the whole universe was nothing but matter relentlessly obeying Newton's laws of motion, future positions and motions were fixed by present positions and motions. And of course the present was completely fixed by the past. Here was the origin of the idea of a completely deterministic universe. Since human bodies and brains were made up of matter and the future state of matter was predetermined by the past then surely it followed that the operation of bodies and brains was predetermined. If mind had to have correspondence with brain activity, how much space was left for free will? Then something happened to blunt the argument. At the atomic level, Quantum Theory emerged and replaced deterministic by probabilistic links.

But what caused the cracks in Newton's universal theory?

What led to absolute time and distance being abandoned as a principle?

What led to the link between cause and effect being weakened?

The answer is light.

Measurement of the speed of light in vacuum showed it to be the same regardless of the speed of the observer relative to the source. Speeds add directly in Newton's classical physics. If you walk along the floor of a moving railway carriage towards the front of the train, then your speed relative to the outside embankment will be your walking speed plus the speed of the train. This doesn't happen with light in a vacuum. Its speed is always measured to be the same (299 million metres per second). Einstein formulated his famous Special Theory of Relativity to solve this dilemma. He proposed that measurements of distance and time intervals are changed by the relative motion of the observer and observed object. These changes make the speed of light seem constant for all observers. Later in his General Theory of Relativity time and distance intervals additionally

limits to intuition

became dependent on the local strength of gravity (which depends on the distribution of mass, the amount of matter). All subsequent experimental observations have so far supported Einstein's Relativity Theories. The passage of time has lost its universal independent status. The observed ticks of any clock, machine or biological, depend on relative motion and gravity. There is no universal time. In practice the size of the deviations from the constant universal time and distances of Newton's model are tiny except at high relative speeds or in very strong gravitational fields. However they are very evident in the more extreme conditions discussed in Chapters 2 and 3, such as close to black holes or in the stretched lifetime of cosmic ray particles produced by collisions in the upper atmosphere. These particles are otherwise too short-lived to reach the surface of Earth. They also become significant at the edge of the observable universe where the velocity of expansion is greatest and is a significant fraction of the speed of light. Relativistic time effects are also seen under much more mundane conditions when ultra-precision measurements are being made. The GPS navigators, in daily use by walkers, sailors, pilots or in our cars can only deliver their incredible accuracy because they calibrate out relativistic effects.

The changes to the notion of cause and effect, also brought about by light are even more difficult to reconcile with our everyday intuition. In chapter 3 we discussed the background microwave radiation, the remnant of an early phase of the universe when matter and radiation interacted intimately in a state of uniform thermal balance. We compared this with the potter's kiln. Both are 'black-body' radiators. The shape of 'black-body' radiation is shown in *(Figure 2.5)*. The classical model of light radiation was unable to satisfactorily account for this shape. The classical model describes light as oscillating electromagnetic waves *(Figure 7.0)*.

Energy is transported by the wave structure in a continuous flow through space. It is easy to see the wave property of light. It shows in the different colours of a soap bubble. The colours are caused by two sets of interfering waves. One is reflected from the outside surface and one from the inside surface of the very thin wall of the bubble. Different wavelengths (colours) are reinforced or cancelled in different places because the bubble wall varies in thickness, changing the relative timing of the reflected waves. This process of wave interferenceis just the same as we see when we drop two stones into a pond and watch the peaks and troughs of the ripples add

limits to intuition

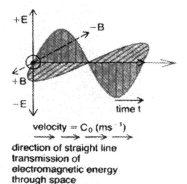

Figure 7.0
Title: **'classical model of light'**

In this model light consists of two coupled waves. An electrical wave (orange) oscillates in strength in the plane of the page. At right angles a magnetic wave (green) oscillates in strength through the page. The two waves reinforce each other and travel together to the right at the speed of light (299 thousand kilometres per second). Energy is distributed continuously throughout the waves. **There are no photons.**

Source: own

and subtract. Light shows wave behaviour but a wave structure alone could not explain the spectrum of blackbody radiation. Plank and Einstein introduced a new solution to the problem. They suggested that light is not only a wave but also a stream of particles. The energy in a light beam is not continuous but in granular packets (eventually named photons). The energy contained in each photon uniquely ties to the wavelength of the wave part of its dual personality. The greater the energy per particle, the shorter is the corresponding wavelength. Many subsequent experiments demonstrated the particle personality of light. Today's instruments are sensitive enough to detect single photons. Modern telescopes take many hours to build images of the faintest galaxies by collecting individual photons.

One problem solved, another created. How can light be both a discrete particle and a continuous wave? The problem didn't stop with light. Experiments with electron particles passing through thin sheets of metal made the same tell-tale wave interference patterns. Particle-wave duality became more general. These issues drove quantum physics. The subject grew in breadth and depth and led to a new theory covering the behaviour of

limits to intuition

light and electrons, Quantum Electro Dynamics (QED). The practical success of QED was enormous and its predictive calculations confirmed by experiment to great accuracy. But the practical success of quantum theory was not matched by any universal agreement on what wave-particle duality really is. Early on an interpretation involving probability gained favour. The wave part of the duality (which is distributed through space just like a normal wave) was taken to control the probability of a particle being at a particular place. In practice the wave is treated just like a classical light wave. The same calculations are done to work out how different waves interfere with each other as they pass through slits or holes or lenses to give a combined wave *(Figure 7.1)*.

In the case of light this resultant 'wave function' is used to predict the probability of there being a photon at any point (it's actually the square of the amplitude of the wave which gives the probability). The wave has no other physical interpretation. There is no way of knowing where any particular photon will go on any occasion. Individual cause and effect relationships are abandoned, probability rules. For interactions involving large numbers of atomic particles and photons, the averaging effect makes such interactions look 'classically' deterministic.

The Conscious Universe

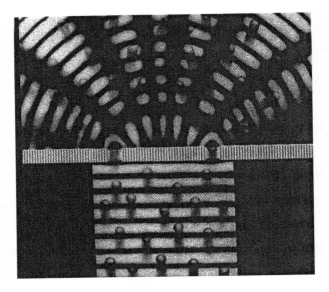

Figure 7.1
Title: **'wave-particle duality'**

An idealised picture of wave-particle duality. In the lower half of the figure the peaks and troughs of the **wave aspect** of light advances upwards towards an opaque screen which has two narrow slits in it. On the tops side of the screen the wave emerges in all directions from each slit (a general property of waves, if the slit is much smaller than the wavelength). Waves from the slits meet and add. Depending on the difference in distance from each slit, they will add at different points on their cycle. This makes the interference pattern running out at right angles to the wave peaks.

The **particle aspect** of light is in the balls representing the photons. The photons pass through the slits and spread out to make their average numbers in any region of space fit with the intensity of the interference pattern established by the waves. This kind of model is used with great success in studying the electron structures within atoms. Philosophical debates on the interpretation of this model still continue nearly one century after its invention.

limits to intuition

There is another important point: the random nature of individual events is taken to be a real feature of the quantum world. For example although the average rate of decay of a radioactive isotope is known, the decay of any particular atom is absolutely random and not dependent on its particular history. However to maintain a precise average individual atoms must operate under similar but unsynchronised propensities to disintegrate. This is in strong contrast to other random events. Perhaps we should call them 'classical' random events. Think of tossing a coin. The reason we can't predict whether it will fall as a head or a tale is because of our lack of detailed knowledge of the exact force we applied when we tossed it. If we knew that its behaviour could be predicted to enormous accuracy using Newton's laws of motion. When the coin left our finger the final outcome was cast. This is so because the coin is an object with so many atoms that the random quantum effects average out. Randomness in the large-scale world is different from randomness in the atomic quantum-world. Randomness in the large-scale world can be about our inability to know enough detail to make a successful prediction, like the balls bouncing around ina lottery-draw machine. They are thoroughly deterministic because of their size but unpredictable because the initial entry state of the balls and other fine mechanical data is in practice unobtainable. This is the territory of Chaos. Chaotic systems can be externally unpredictable but internally deterministic.

Are quantum effects important in brain tissue? The photon structure of light limits the sensitivity of the retina. The human eye can detect small numbers of photons and some animals can 'see' even smaller numbers. Low light levels mean fewer photons. The energy of each photon corresponding to a given colour remains just the same. Are any of the exotic parts of the quantum model below the level of the standard chemical model, involved in neural processing? The brain shows a complex organisation of material engaged in fairly large-scale electrical and chemical processing. Outside of the ultra sensitivity of the eye to low light levels there is no general acceptance of individual quantum events having special significance. But we should be cautious. If evolution produced minimum energy solutions for neurones how far towards the quantum limit would they go? In what way would quantum randomness interplay with larger scale chemistry? But perhaps it's more likely that the quantum level has no practical involvement in the higher operation of our brains.

The quantum world holds many puzzles and has many strange aspects. It has been so since its inception in the first few decades of the twentieth century. In the words of the Nobel Prize winner Richard Feynman, one of the physicist pioneers of QED,
"no one understands quantum theory".
The founders struggled to build an intuitively satisfying picture of what the model said about 'reality'. This was never successfully accomplished. Practical application of the new theory accelerated allowing outstandingly precise predictions. Quantum theory has facilitated massive developments in materials, communications and computing.

This highlights an important distinction between science's power of prediction and its power of explanation.

Testing prediction by experiment and observation is central to model building in science. The philosopher

Karl Popper attempted to capture the essence of scientific method.

His summary was along the lines: to belong to science a theory or model must be expressed in a way which allows experiment and observation to refute it. It should expose itself to testing. No amount of successful testing can ever show it to be absolutely true. It only serves to increase confidence in the theory's present value. Testing can only prove that a theory is definitely false, not that it's definitely true.

Given the passage of time and more evidence old theories inevitably are replaced by newer theories. The old theories may still remain as an approximation of the new theory. This is likely if they are easier to use than the new and if they are accurate enough in practical applications. Newton's theories of gravity and motion have such a life although they have been superseded theoretically by Einstein's General Theory of Relativity. They are still used to design cars, aeroplanes and to plan space flights to the planets.

The 'truth' of a scientific theory or model is assessed through its power to predict a practical event in the 'outer world'. A scientific idea has no more basic measure. This gives it a sort of correspondence with the behaviour of the outer world. Even when we are satisfied with the predictive power of a particular model we still don't have full correspondence to the *workings* of the outer world. The outer world doesn't work by calculating equations. Equations only offer a symbolic metaphor or analogy. The planets travel round the sun without calculating equations.

CHAPTER 8

the unaccounted inner world

Chapter 6 described the screens behind which the 'outer' world is hidden from us. There are further problems in the way of science 'explaining' our experience of feeling and consciousness. The theories and models of the physical sciences are constructed by deliberately ignoring the inner experience of feeling. This has vital consequences.

To illustrate this let us return again to consider the sense of sight. Think about the feeling of seeing the colour we call red. Look at something red. Feel the immediate sensation of redness. When science tries to discover what physical process lies behind the sensation of seeing red, it does not include the actual felt experience of the sensation. Instead it replaces the personal experience with a collection of measurements and ideas. Together these ideas become a theory or model of how red light interacts with our vision system (described in Chapter 5). In chapter 7 we outlined how light can sometimes be described as a coupled vibration in an electrical and magnetic field. The frequency (or wavelength) of the vibration determines which colour we see. Other times it is more useful to describe light as a stream of particles of energy (photons) where the energy of the particles determines which colour we see. However, neither describes what we feel.

The intent of physical science is to construct theories or models which can predict the behaviour of the physical world. To make a prediction, a current or past observation has to be entered into the model. This sets its starting condition. The model is then calculated forward to produce a prediction. However the 'sensation' felt during any observation doesn't have a place in the theory or model. It is discarded. Only numbers distilled out of measurements go forward into the model.

the unaccounted inner world

Let's illustrate this point further. Consider an astronomer who during a series of observations discovered a new planet. He could record the numbers which expressed the planet's observed positions. These could be entered into a mathematical model to predict the planet's position at some time in the future. However nowhere in his calculation would there be a term to capture his conscious experience of 'seeing' the planet. Calculation would enable him to point his telescope at a specified position in the future. If the model was 'correct' he would experience the sensation of the event which corresponded to the prediction. But the internal workings of the model would contain no reference to the 'conscious' quality of either the first or the predicted observation. Physical science makes no attempt at including the experience of consciousness within its models. It has developed its 'objective' position by this deliberate neglect.

There is a fundamental difference between predicting future sensations and explaining how sensations can occur at all.

We can take the sight example again. Above we outlined the process physical science uses in dealing with what we experience as seeing the colour red. This amounts to the substitution of the inner feeling of seeing red by a description of red light in terms of the wavelength of an oscillating electromagnetic field (or the energy of the individual photons in a stream). These models do not explain nor even describe the feeling of seeing red.

However these scientific models do allow the prediction of the behaviour of light in 'the outer world'. They allow future states of wavelength or energy to be calculated with astounding accuracy. These predictions are tested by observation of the real (non-model) world. A prediction is regarded as successful if a sensation is felt in the 'inner world' of the observer at the place and time of the 'external event' predicted by the theory or model. The sensation in the observer may be as 'direct' as seeing red light or as indirect as seeing predicted readings on an instrument or computer printout.

The theory or model does not include this new felt experience of confirmation anymore than it included the 'conscious' aspect of seeing the colour red.

the unaccounted inner world

Let's use another more everyday illustration (*Figure 8.0*).

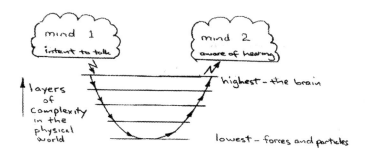

Figure 8.0
Title: **'self to self'**

        The many levels of conversation.

Source: own

A person has an idea which they want to communicate to a friend. The first effect of this idea (which can be seen by others) shows up as electrical activity in this person's brain. In turn this electrical activity is converted by muscles in the larynx into the movement of molecules of air. Vibrating air pressure waves spread from the talker's mouth. Now the reverse process occurs. The vibrating waves pass into the listener's ear causing their eardrum to move. This movement is converted back into electrical activity by the inner ear and transmitted into the listener's brain.

Scientific modelling is very successful in describing all the physical processes in this example, connecting brain stuff to brain stuff. However the trail seems to disappear at each end. It emerges from somewhere else, as a conscious sensation of an idea. It then it dissolves back into the same domain in the mind of the listener. Again it becomes the conscious sensation of an idea.

This notion of existence at several levels attends any system which has coding or representation. Consider all the scientific models describing the path from transmitting brain-stuff to receiving brain-stuff in *(Figure8.0)*. There will some models dealing with molecules and ions and others dealing with networks of cells. At one level all these models are simply differently shaped marks on a piece of paper. The shapes we think of as symbolic script in a natural, mathematical or chemical language. Through the visual system these symbolic shapes cause electrical and chemical activity in the brain-stuff of the scientist looking at them.

the unaccounted inner world

Standing on top of all this, at another 'level' altogether, is the scientist's conscious awareness of what he or she is reading. Worlds upon worlds!

All of this leads us to the compelling question......

Is science capable of explaining where our conscious experience of sensations and thoughts comes from?

Somehow the regions of space occupied by brain tissue have the capacity to produce something of a different nature, a conscious feeling.

From within our own isolated region we each look out and try to make sense of it all.

Taking account of all the topics of the previous chapters we will propose a model to help review our position. There are three parallel worlds to consider.

1) The inner world
2) The outer world
3) The scientific world

Let us describe each of these in turn......

1) The 'inner world' is the world we feel and live in everyday as conscious beings. This is our common-

sense world. We can describe this as an inner world because it is made by our 'brain/mind' from the last link in a chain of physical events which connect us to the second world, the 'outer world'. This last link shows up as a pattern of brain activity which is described in terms of electrochemistry and network interconnections. Currently physical science sees nothing else observable between this activity and our experience of feeling.

2) The 'outer world' is the world which is there without us. It would still be there if there was no conscious life in the universe. The outer world is changed when we make things. Our history is our interaction with the outer world. The outer world doesn't connect inwards to our inner world except through the last links from our sensory organs into the activity of our 'brain/mind'.

3) The third world is the 'science world' which humans have constructed. It is a collection of ideas, theories and models. It is an attempt to 'explain' as much of our world as we can. It is a shared construction capable of third party verification by its own rules of

the unaccounted inner world

evidence. This drive for impersonal objectivity is the historical strength of science. It is also a reason for it to be at its weakest when facing an entirely subjective experience.

Common sense doesn't distinguish between 1) and 2), the inner and the outer worlds. For everyday purposes 3), the science world is taken to be the same as 2), the outer world. This works well in practice. Indeed science predicts the physical behaviour of most of the outer world astoundingly well. This success continues to grow with time across a wide range of specialisms. It's only when we start thinking about the distinctive properties of the inner world of consciousness and feeling that serious limitations arise. These are perhaps not just the limitations of today's scientific knowledge but limitations in principle.

(Figure 8..1) Illustrates the interconnection of these three worlds. The effects of the outer world pass through our senses into the inner world where they become our conscious experience. This chain of processes is where the inevitable preconditioning of our view occurs, fixing how the outer world will look to us. Our inner world takes us further. We build ideas which try to 'explain' our experience. Science is one outcome of this endeavour. We design scientific experiments in our inner world which change parts of the outer world. But again we cannot see the changes except through our senses. (Figure 8.1) shows the complete loop made by this activity.

All of our ideas are based on the experiences of our inner world. Because of this fact, it is probably impossible to explain the existence of the inner-world property of conscious feeling. The explanations become circular. Our scientific theory of conscious feeling will inevitably use ideas which have come from our ability to feel. Perhaps it's like the problem of lifting ourselves off the floor by pulling on our shoelaces. More training in the gym will not solve the difficulty!

the unaccounted inner world

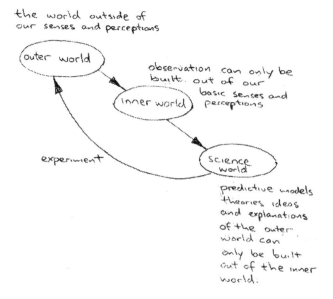

Figure 8.1
Title: **'three worlds interact'**

Our 'inner' world only sees the 'outer' world through filters and preconditions. We use the material of our 'inner' world to make a world which parallels the 'outer world' using scientific models and theories. They have tremendous predictive powers but are human conceptions and cannot capture the full reality of the 'outer world'.

Source: own

In everyday life we constantly experience the sensation of being surrounded by a world of touch, colour, taste, smell and sound. We take it to be exactly what we experience it to be. We consider our experience to be identical with its 'outer world' cause.

Science shows that other processes lie behind these everyday experiences. It gives names and descriptions to these 'behind the scenes' entities. We extend the same feeling of reality to these entities as to those which evolution has placed in our common-sense. So we talk about atoms and gravity in the same way as we talk about apples.

But what is the deeper relationship between the scientific theories and models we carry in our minds and the 'outer world'? Do the ideas of science correspond to things actually existing in the 'outer' world? Not all of them do. Let's illustrate this with a familiar example, the daily weather.

The term atmospheric pressure is commonplace. When the term was first applied to gases, there was no real understanding of what caused pressure. It was defined in practice by its gross effect. Pressure was simply force per unit area. A deeper explanation came later. Gas pressure is caused by molecules in motion, bouncing off each other and off any other objects present in the gas. The 'high-level' concept of pressure was seen to be just the averaged effect of numerous 'lower-level' events. Pressure was simply lots of individual molecular collisions.

the unaccounted inner world

Although this has been fully understood for centuries the averaging idea of gas pressure is still retained. This is because it makes calculation possible. Calculating for all the individual molecules needs impossible amounts of computation

However the most direct model of what is happening in the atmosphere doesn't need the concept of pressure. It needs nothing other than the forces between the particles of the atmosphere and between the particles of the atmosphere and the rest of the universe. Gas pressure is a useful term in our science world but it's probably not needed in the 'outer world'!

If we push this question to the limit we could ask whether the 'outer' world ever runs using anything corresponding to a high level scientific idea (like pressure in the example above)? Or does it work only through the most basic forces always from the bottom-up? Or even in some entirely different way which we have no sight off?

The 'truth' of a scientific theory or model (how closely it corresponds with the outer world) is assessed through its ability to predict future experience. This test gives it a sort of surface correspondence with the 'outer' world but not a guarantee of anything deeper.

Let's look more closely at this question of the correspondence. What is the parallel between a scientific theory and how the outer world actually works. Newton's theory of gravity, dealing with planets and apples can serve as an illustration.

This model is expressed in the equation…..

$$F = \frac{G * M1 * M2}{D^2}$$

where 'F' is the force between two objects with masses of 'M1' and 'M2' which are distance 'D' apart and 'G' is a number which remains constant, initially chosen to suit whichever units of force, mass and time are being used.

This mathematical-model works extremely well over a huge range of circumstances. It allows the calculation of forces which can then be used within other mathematical equations of motion to predict the future motion of objects such as the planets around the Sun. This particular theory has been replaced by Einstein's General Theory of Relativity which allows prediction over an even wider range of conditions.

That changes nothing. The 'nature' of the newer theory is just the same. It is a better model but still a model

the unaccounted inner world

expressed in mathematical symbols. Although Newton's model gives excellent predictions about the behaviour of masses like planet Earth or an apple, it gives no insight as to what the force is or how the force works. The formalism of Einstein's model is conceptually different. In his model the presence of mass changes the shape of space. Objects travel in curved lines because space curves near matter. Gravitational force is not needed as it is in Newton's model. But Einstein's theory is inevitably just as quiet on how nature actually curves space as Newton's is on how nature actually makes gravitational force. In essence the testable aspect of either theory rests on how successfully they can connect one set of measurements to another.

Both are just like a recipe which tells you the temperature and time settings for the oven but says nothing about how food actually cooks.

The Nobel laureate physicist Richard Feynman often emphasised this point. It bothered him that it would take a computer an infinite number of steps to completely run the contemporary model of even the tiniest region of space and time. Nature just wasn't doing it that way. Nature simply doesn't do mathematics in the sense of processing numbers. The equations of physical science can be seen as the equations of models which can be made of mathematical materials. They are not a replacement for reality but a partial analogy or metaphor.

Science endeavours to stack its models in a hierarchy. Models of more complex behaviour stand on top of models dealing with less complex behaviour. Genetics, biology and biochemistry build on top of more basic chemistry. Basic chemistry is built on the physics of QED. This is built on the quantum model of more elementary particles and forces. So the trail goes back to fundamental physics.

As we discussed at the beginning of this chapter all of the models in all of the layers of this hierarchy are made from terms and formulae which have no place to include any aspect of the experience of feeling. There is an act of faith implicit in using these models to try to explain the experience of feeling. It is that the 'explanation' of feeling will selfconstruct out of this unlikely material. For any model to do this it would have to reach out to a level above itself. Surely this is impossible.

A model cannot be our whole world. But our feeling experience of a model is a complete part of our world!

the unaccounted inner world

Any physical model seeking to explain the existence of conscious feeling will have to face these issues. It seems unlikely that more science of the same kind can solve this enigma.

These are very difficult things to comprehend. The more we think about it all, the stranger it seems.

CHAPTER 9

conscious machines

Let's look at this question of conscious feeling from a different direction. We will take our inner world out of the loop as described in *(Figure8.1)*. In its place we will put a machine which embodies some of the predictive models from the world of science. Let us make this an 'artificial' learning machine. It will have its own experimental interactions with the outer world and continuously improve its performance. This will be a useful thought experiment. First a little background.

Over the past 40 years the research field of artificial intelligence or AI has greatly expanded. Two once separate strands began to weave together in the late 1980s. These two strands are similar to two human mental abilities. First is the ability to reason in terms of language. Second is the ability to recognise a visual pattern such as the face of a friend in a crowd.

1) Reasoning in language.

Machine reasoning using language attempts to mimic some of our abilities. Examples include reasoning from the general to the particular and the opposite, seeking the general case from a set of examples. Machine reasoning is still extremely primitive compared with human abilities. A most simple example is a collection of rules of the form......

IF (a particular *premise* is true) THEN (*something else is also true*)

Lots of these rules can be chained together to allow a conclusion to be drawn from many facts. This is the structure of many so-called 'expert systems'. Human experts may write the 'IF...THEN' statements. But much more sophisticated systems have been developed. Systems where the machine can develop rules and their interconnections based on previous example problems with known outcomes. These systems can be initially 'trained' and then left to 'learn' from experience. They have been successful in fields of specialist knowledge which use a limited vocabulary and a limited set of concepts.

2) Pattern recognition.

This is a different approach to a different class of problems. It is most easily illustrated when the patterns are visual, for example when the patterns are fingerprints or faces. The first stage is to select the most distinctive features in the type of image to be scrutinised. This may involve the judgement of an expert or sometimes it may be possible to use an automated search. The chosen features are then measured. These measurements are made the inputs to a machine called a neural network. It is so called because it looks like a gross simplification of a network of biological neurones (*Figure9.0*)

The 'knowledge' content of the network is contained within the scaling factors. The values of the scaling factors are adjusted by training. They enable the network to function as a learning device. During training different known images are shown to the system. The system is 'told' what its decision should be. It then adjusts the values of all its scaling factors to give the correct decision. How it does this is key to the success of the system. The mathematics is designed to calculate a combination of scaling factor values which also gives good rejection of all the incorrect images it has already been shown.

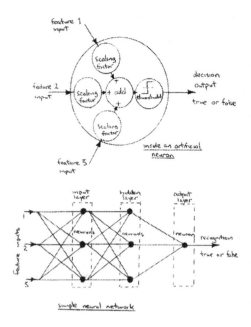

Figure 9.0
Title: 'artificial neural networks'

    Inside each neurone the strength of each input feature is adjusted by a scaling factor before the features are added together. The values of the scaling factors are derived during a training session for the complete neural network. The network is trained on many images to optimise its performance. Within each neurone the sum of the scaled features is compared with a threshold. This comparison decides which of two levels will appear at the neurone's output. At the final output of the network these two levels correspond to recognition or non-recognition. The simple network shown in the lower sketch only has three input features. For more features each layer will have more neurones in parallel. When trained the 'knowledge' held in the neural network is distributed throughout the values of the scaling factors. This is the 'knowledge' the network used to make its decision. However because of its form it is difficult to interpret as the intermediate stage in making the overall decision.

Source: own

Current AI systems may combine language reasoning and pattern recognition in a complementary approach to problem solving. This hints at our own combination of linguistic and spatial skills.

Let's take this general kind of technology and add lots of imagination in order to conduct a thought experiment. We need lots of imagination because the technology we have summarised only performs well in very narrow applications. This technology seeks to mimic behaviour we would regard as 'intelligent' if performed by a human. However, our thought experiment will take us to a different property, that of conscious feeling. Researchers working in AI seek to make machines more 'intelligent' but most of them would balk at the notion of a machine able to feel. However we will suspend disbelief so that we can discuss what might be learned 'in principle' from a conscious machine.

With the qualifications given above in mind, we are ready to build our artificial cat. When interacting with the world our cat will make modifications to its own software and constantly improve those programs which best match 'real' cat behaviour. It will start off getting its walking gait correct. Then tail flicking, mewing, licking and ankle rubbing with an erect tail. The correct shape and texture will have already been crafted into the robot. By now many people will be fooled. But some kind person will start feeding the cat. In order to improve robo-cat we will have to make it eat fish and purr and run to a plate when it is rattled. We will need the latest processors and lots of memory. This adaptive 'intelligent' machine will grow to have the very best of artificial intelligence technology at its core. It will have combined pattern recognition and sequential reasoning and everything new! In the end everyone will be fooled, including the cat next door.

But does our technological cat taste and enjoy its fish?

(The sceptic might say we don't know whether a real cat tastes and enjoys its fish. Let's relax our scepticism for the present and rely on the conviction of cat owners.)

Even if our robo-cat became advanced enough to describe his fish, we could never know whether or not we were just listening to a sophisticated machine with the correct external responses. This line of argument is not new and does not stop with machines. We have no access to other people's feelings. We can only infer they have feelings from our own experience and the circumstantial evidence of what they say how they sound and how they look. However this level of scepticism is rare outside of philosophy books. Even philosophers give it up on their way home in the evening.

Could a machine be built which had or could develop conscious feelings? Is the type of material in brains special? If other types of material can be made to interact in logically equivalent ways then why does the type of material matter? Is matter organised by software, at least in principle, able to replicate the property of feeling. Imagine we start with a biological brain which has conscious feeling. We isolate a small region and identify its electrical and chemical exchanges with the surrounding tissue. Then we make some device which replicates the behaviour of this small region and connect it in as a replacement part. Would this change the person's ability to feel? If not, could we go on and replace more and more. Is biological material any different to pre-biological material? Is it just about complexity in the interconnections of material?

If we suppose that an aware machine can be made then we have to face the likelihood that we would never be able to 'explain' its ability. We don't know how biological brains can feel. We don't know anything to design into a machine. If conscious machines are to emerge they will have to 'evolve' through interaction and adaptation with their environment. This is just like our own brain. We would find nothing in the models we had of such a machine's component pieces or in the models we could make of its massive interconnections which would explain the arrival of feeling.

If conscious feeling is possible for a machine then it will likely be for the same reason that it is possible for a 'biological' brain. The problem is that both cases would seem equally impossible to model. It is hard for us to see how a collection of inanimate components could be connected to become aware, to feel. To illustrate this imagine replicating the feeling of pain. Pain felt within the structure of a machine. At its most basic level the machine is simply moving electronic charge around inside pieces of silicon. Do only groupings of charge movements have the 'feeling' property? Or does it happen at a 'higher' level? Does it emerge up at the level corresponding to symbolic constructions in the self-constructing programs?

This is precisely the same problem as we have to face with the electrical activity of the atoms and molecules in our brain.

There are new interdisciplinary projects trying to address these difficult questions with open ideas. Biologists, mathematicians, engineers, physicists, computer scientists and others are collaborating to study what has become known as 'emergent phenomena'. Emergence is a term increasingly used to describe the arrival of high level behaviour in systems built of low level parts: parts too simple to contain the behaviour. Biological evolution is the grandest such system with molecules as its lowest level parts. Computer models can be made to modify themselves in a process analogous to the mutation and selection of biological evolution. Such models develop properties easily spotted at the top level but difficult to explain from the bottom. In that way they become similar to natural biological systems.

This new direction moves the focus away from material and on to adaptive algorithm. That is to sets of rules which govern relationships. Not simple rules but rules which are conditional and interconnected: rules which self-modify according to their success or failure. The above ties into the techniques of artificial intelligence we outlined above. But intelligence, artificial or otherwise is not the central issue. A dumb machine which could feel would be a miracle compared with the most intelligent machine which doesn't feel anything.

Is the focus on algorithm sufficiently different to escape the problem of explaining feeling? Perhaps more likely as in the case of our cat, if a complex self-adapting system convinces us that it has internal feelings, it will be just as impenetrable to understanding as the natural phenomenon it was meant to explain. Another 'catch 22'? Success resulting in a new phenomenon to be explained as part of the original problem.

CHAPTER 10

taking stock

We live our lives taking so many things for granted. Our mind is amazingly over capable. We could live our lives in a robotic way, responding to our environment. We could search it with 'intelligence', the intelligence of a super-sophisticated bunch of sensors connected to a super-capable computer. We could win our evolutionary way by sheer performance, competing in the environment, making the best cost-benefit analysis to win the survival tests of evolution. Still we need not have conscious feeling. Consciousness and feeling are beyond all of this. When we raise the level of consciousness above the senses such as sight or touch and reach the sublime of music or poetry, we are where there is no mechanistic reason to be.

Our journey has been like climbing the spiral staircase of a lighthouse. Each new turn gives us a new window. When we look out we see the same foreground but as we go higher, we see just a little bit further. What is our present view? Science is incredibly successful in describing the behaviour of the 'outer' world. It's predictive, manipulative and explanatory powers are immense. It's when we try to explain our 'inner' world that science has difficulties.

The content of physical science is a world devoid of feeling. It tries to build a picture of phenomena while excluding the feeling perception of the observer. Only bare instrument readings or other forms of objective data are taken forward. These are packaged into a theory or model. The model is validated by the way it can predict future instrument readings or events in further observations and experiments.

Physical science builds theories and models from observations and experiences measured (and felt) by scientists but removes any reference to the experience of feeling. The subjective experience of feeling is never retained in any parameter of the model. The essence of science is to exclude the subjective experience of phenomena and replace them with the subjective experience of a theory or model. A felt experience is replaced by a model which contains no reference to feeling, a model that is of necessity, a level less than full blown experience.

Does this lead to the inevitable incompleteness of any scientific attempt to explain our conscious selves? An analogy might help. Our scientific ideas about the substance of the universe can be considered as being like shadows. Certain dimensions are preserved in a shadow but some are lost. We can find consistent relationships between the shadows. These relationships tell us things about the full three-dimensional object making the shadow. Consciousness and feeling lie in the full substance of reality. Science works on the shadows, with fewer dimensions. Never the less the shadows of science are far too successful to be mere fictions. They are genuine but limited projections of the outer world. Perhaps we can take the analogy further. The shadow is connected to its physical object because they are both in the same world of geometrical space. But what if feelings are more like the colour of an object, a property completely unable to cast a shadow?

Physical science tries to build models from the bottom up. The models of physics are taken to be fundamental. How is this justified? Why not put all the areas of science in a flatter perspective without taking one to be more fundamental than another? One good reason for physics being considered fundamental is that the things which physics talks about were present earlier than those which lie further up the chain of complexity. The early universe only had the forces and particles of physics. The objects which chemistry studies, other than helium and hydrogen only arrived after the physics of stars had built the atomic nuclei. Chemistry rich enough to do coding only became possible when there were places with suitable chemical environments for complex molecules to survive. Life only evolved when complexity developed through the type of 'self-design' we outlined in Chapter 6.

Higher levels do change lower levels in an environmental sense but not to the point of changing the behaviour of the fundamental forces of physics.

Darwin's Theory of Evolution was forced to work from the top down. The underlying science was not then available. Attempts to draw conclusions from top-level features across species or within species are tentative and

taking stock

open to many schemes of accounting. They have a high ratio of opinion to evidence. They are more descriptive than predictive. They are statistical rather than causal.

With advances in biochemistry, cell biology and genetics it has become possible to

delve down and uncover the detail. The detail expands to reveal layer upon layer of further detail. Mechanism everywhere but all based on the way molecules interact. Rising in complexity, no chemistry is broken supporting the higher level mechanisms. However the higher mechanisms are too complex to predict from the bottom. They can only be revealed from the top.

Bottom-up accounts, where they are possible, are much stronger and better connected to less ambiguous data. They are more evidence driven. Bottom-up accounts through biochemistry, cell biology and genetics are getting stronger and stronger. With time the frontier is moving up the hierarchy. But it has a very long way to rise on the scale of life's complexity.

It is likely that all sorts of unseen mechanisms exist in the way DNA maps into the outcomes of the living world. The notion of 'self-design' outlined in chapter 6 may have many levels. The coding schemes themselves will have many built-in aids to self-design. There must be huge amounts of intermediate mechanism hiding behind the curt terms 'mutation' and 'natural selection'. However in the end there has to be consistency throughout the bottom-up chemical thread, even when it weaves into the incredibly complex world of which we are a part.

Suppose we assume that eventually after enormous amounts of effort, dedication and time all of the levels of chemistry are connected from the bottom upward and we feel more secure in our understanding of the 'inside' of the evolutionary process. Where are we then? We still have a problem of 'explanation' but it is more obscure. We still have the end problem of relating brain to mind, chemistry to feeling and consciousness. The top end of complexity leads to an enigma which is outside of the system of explanation in terms of biochemistry, biology, genetics and emergent systems. If we accept we have a problem here we just can't ignore it and carry on assuming it has no relevance in other arenas of 'explanation'. This seems important. It should make us sceptical about taking as complete any of our fundamental concepts about the material world. It's not easy to accept that low-level atomic forces, alone, have the possibility of achieving so much.

Is it impossible to ever fully 'explain' consciousness (for the fundamental reasons of filtering, self-reference and omission reviewed in chapters 6 and 8)? Labelling it as an 'emergent phenomena' seems no more than a description. Consciousness is a working part of the loop of all science. We should therefore be sceptical about the completeness of any scientific 'explanation'. This reservation becomes strongest when 'explanation' is targeted at the level of feeling and consciousness.

The search for underlying mechanism does pay off. We see and connect more and more. All seems well until we look at the 'top-level' of ourselves. How does any of this wondrous mechanism of molecules, however complex, produce the experience of feeling?

Is feeling in the molecules? No!

Is feeling in the bonds making the molecules into ribbons and spirals and balls? No!

Is feeling in the electrical charge moving between the molecules? No!

Is feeling in the 'organisation' of the whole system? This is the current answer for some.

But what is this notion of organisation? Is it anymore than the previous list of entities? Is the notion of a separate entity 'organisation' not simply an abstract idea only able to exist inside a mind? Again this is a property trying to explain itself, in terms of itself. We can see all the physical correlations in our brain but not the link to the experience of feeling.

The gap in explaining our feeling selves, which seems to be inevitably left by science, can be filled by individual persuasion involving an act of faith. Alternatively it can be left unfilled, as in the agnostic sense. However if it is filled there should surely be some compatibility with the kind of scientific knowledge we have been discussing. More knowledge born out of profound curiosity and observation must illuminate a worthwhile view of the world.

But we have our fears and our past heritage of ideas to contend with.

taking stock

What is satisfying to us,

(a) a continuous sense of amazement about our universe and a belief that our conscious self ends here on planet Earth, when we die?

(b) a feeling that this is only a transition for our conscious self, to something else?

Belief (a) is easily sustainable if we just look at the mechanisms of our world.

Belief (b) is difficult to sustain if we just look at the mechanisms of our world. To hold such a belief requires an act of faith.

We have spent a lot of time looking at the intimacy of correlation between feelings and the material structures of our brain (all the way down through the underlying chemistry and physics). In all of this we have seen that the 'emergence' of feeling, correlates with increasing complexity in the organisation of material structures. This is true whether we view the long historical path of evolution or the 'fast track' path of the developing embryo. Having seen the correlation on the up-slope of complex organisation, we should consider the downslope. What happens when the material structure comes apart and returns to a state of disorganised material? What happens when our brain dies and dissolves back into its basic chemical constituents? What becomes of conscious feeling, of self?

The intimate correlation between conscious feeling and the material structure of our brain makes it difficult to understand the continuing existence of conscious feeling without its material substructure. The possible independent existence of feeling seems to clash with a common-sense response to our current knowledge.

There are obviously other ways of 'going on', in a sense, beyond our mortal span within which we can find satisfaction. We can 'go on' through our influence on other people who live beyond our life span. We can build things, produce art, produce ideas, produce values, show good example. Regardless of our agnostic or other position we find satisfaction in the development of human culture: the ability to transmit a collective heritage onwards and if we are lucky, add an atom to it.

Our purpose has been to point out the inevitable gap in our understanding. The agnostic position would leave the gap unfilled. The traditional religious position would fill it with a 'different' type of knowledge, supported by an act of faith.

A further level of difficulty can arise here. Ideas about how our material world works can become bound together with rules for living, ethical and moral codes. What happens to our consciousness after death is linked to an 'afterlife' or the 'next life'. If we accept that the evidence does not support the conscious self outliving the material brain then hopes for our personal survival in an afterlife collapse. But that potential loss should not be allowed to colour our picture of the physical world.

Many individuals show how science and faith can coexist. Success seems possible when faith operates on the highest level of the 'why' questions and is not prescriptive about how the natural world works.

In earlier chapters we asked what corresponds in the 'outer' world to the hierarchical levels of our explanations. Do such hierarchical divisions only exist inside the models we build in our mind? Perhaps we could view religious ideas, developed through history from origins in prehistory, as model building aimed at the top level of cause and purpose. We can then ask as before, where this top-level hierarchical entity actually lives, only in minds, or somewhere else? A personal answer to the time old question 'does God exist' could be yes but perhaps only inside mind and only from there, able to effect our material world. Has mind discovered or invented a higher example of itself?

Each individual's faith is their decision and by definition can never be proved. Every individual should be free to choose their worldview. A view hopefully nurtured from as much available information as time will permit and never by the deliberate exclusion of hard won knowledge about the practical workings of our universe. There is also a sheer delight in trying to see the intricate interconnectedness of everything. As a practical subject science has provided the most amazing insights into our world. It is more wonderfully constructed than any imaginings. Science illuminates our worldview and expands our sense of wonder and awe.

This is what we should nurture from science, the expansion not the contraction of our spiritual mood.

POST SCRIPT

Below are two paragraphs quoted from Erich Heller. They seem to resonate with the ideas of this book. Maybe you will feel the same, if you will allow 'consciousness' to have its poetic outing as 'spirit'.

"Spirit is the traveller, passes now through the realm of man. We did not create spirit, do not possess it, cannot define it, are but the bearers. We take it up from unmourned and forgotten forms, carry it through our span, will pass it on, enlarged or diminished, to those who follow. Spirit is the voyager, man is the vessel."

"From primal mist of matter to spiralled galaxies and clockwork solar systems, from molten rock to an earth of air and land and water, from heaviness to lightness to

life, sensation to perception, memory to consciousness.... man now holds a mirror, spirit sees itself. Within the river currents turn back, eddies whirl. The river itself falters, disappears, emerges, *moves on. The general course is the growth of form, increasing awareness, matter to mind to consciousness. The harmony of man and nature is to be found in continuing this journey along its ancient course towards greater freedom and awareness."*

FURTHER READING LIST

Malcolm S. Longair. *Our Evolving Universe*. United Kingdom. Cambridge University Press, 1996,
ISBN 0 521 55091 2

Tom Wilkie & Mark Rosselli. *Visions of Heaven*. London, Hodder & Stoughton, 1998,
ISBN 0 340 71734 3

John D. Barrow & Joseph Silk. *The Left Hand of Creation*. London: William Heinemann Ltd., 1984,
ISBN 0 04 523002 1

Barrie W. Jones, Robert J. A. Lambourne, David A. Rothery. *Images of the Cosmos*. United Kingdom, The Open University, 1994,
ISBN 0 340 60065 9

Richard P. Feynman, *QED the strange story of light and matter*, Penguin Books Ltd, England, 1990,
ISBN 0 14 012505 1

Richard P. Feynman, *The Character of Physical Law*, Penguin Books Ltd, England, 1992,
ISBN 0 14 017505 9

Sir Arthur S. Eddington. *The Nature of the Physical World*. London, J. M. Dent & Sons Ltd., 1935

Steve Jones & Borin Van Loon, *Genetics for Beginners*, Icon Books Ltd., Cambridge, 1993,
ISBN 1 874166 12 9

Ian Wilmut, Keith Campbell, Colin Tudge, *The Second Creation*, Headline Book Publishing, London, 2000
ISBN 0 7472 5930 5

John Maynard Smith & Eors Szathmary, *The Origins of Life*, Oxford University Press, New York, 1999,
ISBN 019 850493 4

Michael J Behe, *Darwin's Black Box*, Touchstone, New York, 1998
ISBN 0 68 483493 6

Richard Dawkins. *The Blind Watchmaker*. London, Penguin Books Ltd., 1988,
ISBN 0 14 014481 1

Richard Dawkins. *Unweaving the Rainbow*. London, Penguin Books Ltd., 1988,
ISBN 0 14 014481 1

Daniel C. Dennett. *Darwin's Dangerous Idea*. London, Penguin Books Ltd.,1996,
ISBN 0 14 016734 X

Neil A. Campbell. *Biology*, third edition, The Benjamin / Cummings Publishing Company, Inc., California 94065, 1993,
ISBN 0 8053 1880 1

Douglas J. Futuyma. *Evolutionary Biology*, Sinauer Associates Inc., Massachusetts, 1998
ISBN 0 87893 189 9

Anthony J. F. Griffiths and others, *An introduction to Genetic Analysis*, 6th edition, W. H. Freeman and Company, New York, 1996,
ISBN 0 7176 2604 1

Christopher K. Mathews, K. E. van Holde, Kevin G. Ahern, *Biochemistry*, 3rd edition, Addison Wesley Longman Inc., San Francisco CA94111, 2000,
ISBN 0 8053 3066 6

Natural History Museum, *Man's place in evolution*, Cambridge University Press, 2nd edition 1991,
ISBN 0 521 40864 4

John L. Casti. *Paradigms Lost*. London: Abacus, 1990,
ISBN 0 349 10544 8

A.G. Cairns-Smith. *The Life Puzzle*, Edinburgh, Oliver & Boyd, 1971,
ISBN 0 05 002297 0

A.G. Cairns-Smith. *Genetic takeover*, London, Cambridge University Press, 1982,
ISBN 0 521 23312 7

Matt Ridley, Genome. London, Fourth Estate Ltd., 1999,
ISBN 1 857 02 835

Angus Gellaty, Oscar Zarate, *Introducing Mind & Brain*, Icon Books, Cambridge, United Kingdom, 1999,
ISBN 1 8404 60849

John Nolte. *The Human Brain*, Mosby Inc., 1999,
ISBN 0 8151 8911 7

John Allman, *Evolving Brains*, Scientific American Library, 2000,
ISBN 0 7167 6038 X

Lawrence Weiskrantz, *Consciousness Lost and Found*, Oxford University Press, 1997,
ISBN O 19 852458 7

Patricia Smith Churchland. *Neurophilosophy.* Cambridge, Massachusetts, The MIT Press, 1986,
ISBN 0 262 03116 7

Colin Blakemore & Susan Greenfield (Editors). *Mindwaves*. Oxford U.K., Basil & Blackwell Ltd., 1987, ISBN 0 631 14622 9

Susan Greenfield, *Brain Story*. London, BBC Worldwide Ltd., 2000,
ISBN 0 563 55108 9

Rita Carter, *Mapping the Mind*. London, Phoenix, 2000, ISBN 0 753 81019 0

Steven Rose (editor). *From Brains to Consciousness*. London, Allen Lane, The Penguin Press, 1998,
ISBN 0 713 99167 4

Daniel C. Dennett. *Kinds of Mind*. London: Weidenfeld & Nicolson, The Orion Publishing Group, 1996,
ISBN 0 297 81546 6

Joseph LeDoux. *The Emotional Brain*. London, Phoenix (a division of Orion Books Ltd), 1999
ISBN 0 75380 670 3

Francis Crick. *The Astonishing Hypothesis*. London: Simon & Schuster Ltd., 1994,
ISBN 0 671 71295 0

Paul M. Churchland & Patricia S. Churchland. *On the Contrary*. Cambridge,

Massachusetts; London, England: A Bradford Book, The MIT Press, 1998,
ISBN 0 262 03254 6

A.G. Cairns-Smith. *Evolving the Mind*, Cambridge University Press, 1996,
ISBN 0 521 40220 4

Douglas R. Hofstadter, Daniel C. Dennet, *The Mind's I*, Penguin Books, U.K.,1982,
ISBN 0 14 00 6253 X

John Searle, *Minds, Brains and Science*, Penguin Books, England, 1989,
ISBN 0 14 022867 5

Roger Penrose, *Shadows of the Mind*, Oxford University Press, Great Britain, 1994,
ISBN 0 09 9582211 2

John Maher & Judy Groves, *Chomsky for Beginners*, Icon Books Ltd., Cambridge,1996,
ISBN 1 874166 420

James McGilvray, *Chomsky*, Blackwell Publishers Inc., Malden, MA 02148, USA, 1999,
ISBN 0 7456 1888 X

Noam Chomsky, New Horizons in the Study of Language and Mind, Cambridge University Press, United Kingdom, 2000,
ISBN 0521 65822 5

Karl R. Popper, *Conjectures and Refutations*, Routledge & Kegan Paul Ltd., London, 1972,
ISBN 0 71 006508 6

Karl Popper, *Unended Quest*, Fontana, 1976,
ISBN 0 00 634116 0

Bertrand Russell, *The Problems of Philosophy*, Oxford University Press, Great Britain, 1959, (Home University 1912),
ISBN 0 19 888018 9

CPSIA information can be obtained at www.ICGtesting.com
Printed in the USA
BVOW05s1903250116

434176BV00001B/7/P